CORE FOCUS ON
LINEAR EQUATIONS

Stage 3

SMc curriculum

AUTHORS

Shannon McCaw

Beth Armstrong • Matt McCaw • Sarah Schuhl • Michelle Terry • Scott Valway

COVER PHOTOGRAPH

Empire State Building
Opened on May 1, 1931, the Empire State Building stretches 1,454 feet into the sky. The building sits in the heart of New York City and draws millions of visitors each year to its two observation decks.
©iStockphoto.com/Christopher Penler

Copyright ©2014 by SM^c Curriculum, LLC. All rights reserved. Printed in the United States. This publication is protected by copyright. No part of this publication should be reproduced or transmitted in any form or by any means without prior written consent of the publisher. This includes, but is not limited to, electronic reproduction, storage in a retrieval system, photocopying, recording or broadcasting for distance learning. For information regarding permission(s), write to: Permissions Department.

ISBN: 978-1-938801-76-1

4 5 6 7 8 9 10

ABOUT THE AUTHORS

From left to right: Beth Armstrong, Matt McCaw, Shannon McCaw, Scott Valway, Michelle Terry, Sarah Schuhl

SERIES AUTHOR

Shannon McCaw has been a classroom teacher in the Newberg and Parkrose School Districts in Oregon. She has been trained in Professional Learning Communities, Differentiated Instruction and Critical Friends. Shannon currently works as a consultant with math teachers from over 100 districts around Oregon. Shannon's areas of expertise include the Common Core State Standards, curriculum alignment, assessment best practices and instructional strategies. She has a degree in Mathematics from George Fox University and a Masters of Arts in Secondary Math Education from Colorado College.

CONTRIBUTING AUTHORS & EDITORS

Beth Armstrong has been a classroom teacher in the Beaverton School District in Oregon. She has received training in Talented and Gifted Instruction. She has a Masters in Curriculum and Instruction from Washington State University.

Matt McCaw has been a classroom teacher, math/science TOSA and special education case-manager in several Oregon school districts. Matt has most recently worked as a curriculum developer and math coach for grades 6-8. He is trained in Differentiated Instruction, Professional Learning Communities, Critical Friends Groups and Understanding Poverty. Matt has a Masters of Special Education from Western Oregon University.

Sarah Schuhl has been a classroom teacher, secondary math instructional coach and district-wide K-12 math instructional specialist, most recently in the Centennial School District in Oregon. Sarah currently works as a Solution Tree associate and an educational consultant supporting and challenging teachers in the areas of math instruction and alignment to the Common Core State Standards, common assessments for all subjects and grade levels and professional learning communities. From 2010–2013, Sarah served as a member and chair of the National Council of Teachers of Mathematics editorial panel for their Mathematics Teacher journal. Sarah earned a Masters of Science in Teaching Mathematics from Portland State University.

Michelle Terry has been a classroom teacher in the Estacada and Newberg School Districts in Oregon. Michelle has received training in Professional Learning Communities, Critical Friends, ELL Instructional Strategies, Proficiency-Based Grading and Lesson Design, Power Strategies for Effective Teaching, and Classroom Love and Logic. Michelle has an Interdisciplinary Masters from Western Oregon University. She currently teaches mathematics at Newberg High School.

Scott Valway has been a classroom teacher in the Tigard-Tualatin, Newberg and Parkrose School Districts in Oregon. Scott has been trained in Differentiated Instruction, Professional Learning Communities, Critical Friends, Discovering Algebra, Pre-Advanced Placement, Assessment Writing and Credit by Proficiency. Scott has a Masters of Science in Teaching from Oregon State University. He currently teaches math at Parkrose High School.

COMMON CORE STATE STANDARDS
Grade 8 Overview

The complete set of Common Core State Standards can be found at http://www.corestandards.org/. This book focuses on the highlighted Common Core State Standards shown below.

The Number System

- Know that there are numbers that are not rational, and approximate them by rational numbers.

Expressions and Equations

- Work with radicals and integer exponents.

- Understand the connections between proportional relationships, lines and linear equations.

- Analyze and solve linear equations and pairs of simultaneous linear equations.

Functions

- Define, evaluate and compare functions.

- Use functions to model relationships between quantities.

Geometry

- Understand congruence and similarity using physical models, transparencies or geometry software.

- Understand and apply the Pythagorean Theorem.

- Solve real-world and mathematical problems involving volume of cylinders, cones and spheres.

Statistics and Probability

- Investigate patterns of association in bivariate data.

Mathematical Practices

1. Make sense of problems and persevere in solving them.

2. Reason abstractly and quantitatively.

3. Construct viable arguments and critique the reasoning of others.

4. Model with mathematics.

5. Use appropriate tools strategically.

6. Attend to precision.

7. Look for and make use of structure.

8. Look for and express regularity in repeated reasoning.

CORE FOCUS ON LINEAR EQUATIONS

CONTENTS IN BRIEF

How To Use Your Math Book	VIII
Block 1 Expressions and Equations	1
Block 2 Sequences and Slope	44
Block 3 Using Linear Equations	93
Block 4 Systems of Equations	135
Block 5 Two-Variable Data	181
Acknowledgements	225
English/Spanish Glossary	226
Selected Answers	261
Index	270
Problem-Solving	272
Symbols	273

CORE FOCUS ON LINEAR EQUATIONS

BLOCK 1 ~ EXPRESSIONS AND EQUATIONS

Lesson 1.1	Order of Operations	3
Lesson 1.2	Evaluating Expressions	6
Lesson 1.3	The Distributive Property	10
	Explore! Match Them Up	
Lesson 1.4	Solving One-Step Equations	15
	Explore! Addition and Subtraction Equations	
Lesson 1.5	Solving Two-Step Equations	20
Lesson 1.6	Solving Multi-Step Equations	25
	Explore! Multi-Step Equations	
Lesson 1.7	Solutions to Linear Equations	30
	Explore! What Works?	
Lesson 1.8	Linear Inequalities in One Variable	34
Review	Block 1 ~ Expressions and Equations	39

BLOCK 2 ~ SEQUENCES AND SLOPE

Lesson 2.1	Recursive Routines	46
	Explore! Caloric Recursive Routines	
Lesson 2.2	Linear Plots	51
Lesson 2.3	Recursive Routine Applications	56
	Explore! Saving and Spending	
Lesson 2.4	Rate of Change	61
Lesson 2.5	Recursive Routines to Equations	67
	Explore! Modeling with Equations	
Lesson 2.6	Input-Output Tables from Equations	73
	Explore! Linear Qualities	
Lesson 2.7	Calculating Slope from Graphs	77
Lesson 2.8	The Slope Formula	82
	Explore! Find That Formula	
Review	Block 2 ~ Sequences and Slope	86

BLOCK 3 ~ USING LINEAR EQUATIONS

Lesson 3.1	Graphing Using Slope-Intercept Form	95
Lesson 3.2	Writing Linear Equations for Graphs	100
	Explore! Find The Equation	
Lesson 3.3	Writing Linear Equations from Key Information	106
	Explore! Triangle Lines	

BLOCK 3 ~ CONTINUED

Lesson 3.4	Different Forms of Linear Equations	111
	Explore! One of These Things	
Lesson 3.5	More Graphing Linear Equations	115
	Explore! Match Me	
Lesson 3.6	Graphing Linear Inequalities in Two Variables	120
	Explore! In The Shade	
Lesson 3.7	Introduction to Non-Linear Functions	125
	Explore! Non-Linear Curves	
Review	Block 3 ~ Using Linear Equations	130

BLOCK 4 ~ SYSTEMS OF EQUATIONS

Lesson 4.1	Parallel, Intersecting or the Same Line	137
	Explore! Types of Systems	
Lesson 4.2	Solving Systems by Graphing	141
Lesson 4.3	Solving Systems using Tables	145
	Explore! Larry's Landscaping	
Lesson 4.4	Solving Systems by Substitution	150
	Explore! A Trip on I-70	
Lesson 4.5	Solving Systems using Elimination	154
Lesson 4.6	Choosing the Best Method	158
	Explore! What's Easiest?	
Lesson 4.7	Applications of Systems of Equations	161
	Explore! At the Movies	
Lesson 4.8	Systems of Linear Inequalities	167
Lesson 4.9	Converting Repeating Decimals to Fractions	171
Review	Block 4 ~ Systems of Equations	175

BLOCK 5 ~ TWO-VARIABLE DATA

Lesson 5.1	Scatter Plots and Correlation	183
Lesson 5.2	Predicting with Lines of Best Fit	187
	Explore! Finding a Good Fit	
Lesson 5.3	Five-Number Summaries of Data	192
Lesson 5.4	Q-Points and Lines of Best Fit	197
	Explore! The Wave	
Lesson 5.5	Predicting with Best Fit Equations	203
Lesson 5.6	Using Data and Graphs to Persuade	207
	Explore! Eliminating Bias	
Lesson 5.7	Bivariate Data and Frequency Tables	212
Review	Block 5 ~ Two-Variable Data	219

HOW TO USE YOUR MATH BOOK

Your math book has features that will help you be successful in this course. Use this guide to help you understand how to use this book.

Lesson Target

 Look in this box at the beginning of every lesson to know what you will be learning about in each lesson.

Vocabulary

Each new vocabulary word is printed in **red**. The definition can be found with the word. You can also find the definition of the word in the glossary which is in the back of this book.

Explore!

Some lessons have **EXPLORE!** activities which allow you to discover mathematical concepts. Look for these activities in the Table of Contents and in lessons next to the purple line.

Examples

Examples are useful because they remind you how to work through different types of problems. Look for the word **EXAMPLE** and the green line.

Helpful Hints

Helpful hints and important things to remember can be found in green callout boxes.

Blue Boxes

A blue box holds important information or a process that will be used in that lesson. Not every lesson has a blue box.

 This calculator icon will appear in Lessons and Exercises where a calculator is needed. Your teacher may want you to use your calculator at other times, too. If you are unsure, make sure to ask if it is the right time to use it.

EXERCISES

The **EXERCISES** are a place for you to find practice problems to determine if you understand the lesson's target. You can find selected answers in the back of this book so you can check your progress.

REVIEW

The **REVIEW** provides a set of problems for you to practice concepts you have already learned in this book. The **REVIEW** follows the **EXERCISES** in each lesson. There is also a **REVIEW** section at the end of each Block.

TIC-TAC-TOE ACTIVITIES

Each Block has a Tic-Tac-Toe board at the beginning with activities that extend beyond the Common Core State Standards. The Tic-Tac-Toe activities described on the board can be found throughout each Block in yellow boxes.

CAREER FOCUS

At the end of each Block, you will find an autobiography of an individual. Each one describes what they like about their job and how math is used in their career.

CORE FOCUS ON MATH
STAGE 3

 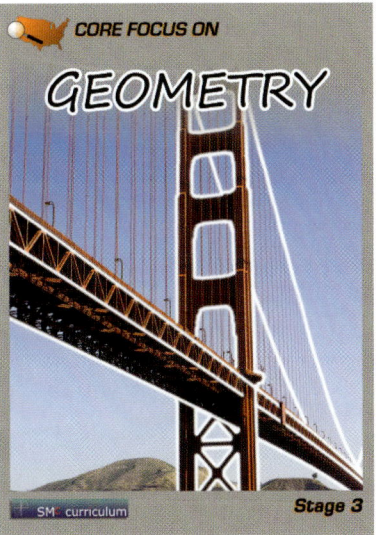 

CORE FOCUS ON LINEAR EQUATIONS
BLOCK 1 ~ EXPRESSIONS AND EQUATIONS

Lesson 1.1	Order of Operations	3
Lesson 1.2	Evaluating Expressions	6
Lesson 1.3	The Distributive Property	10
	Explore! Match Them Up	
Lesson 1.4	Solving One-Step Equations	15
	Explore! Addition and Subtraction Equations	
Lesson 1.5	Solving Two-Step Equations	20
Lesson 1.6	Solving Multi-Step Equations	25
	Explore! Multi-Step Equations	
Lesson 1.7	Solutions to Linear Equations	30
	Explore! What Works?	
Lesson 1.8	Linear Inequalities in One Variable	34
Review	Block 1 ~ Expressions and Equations	39

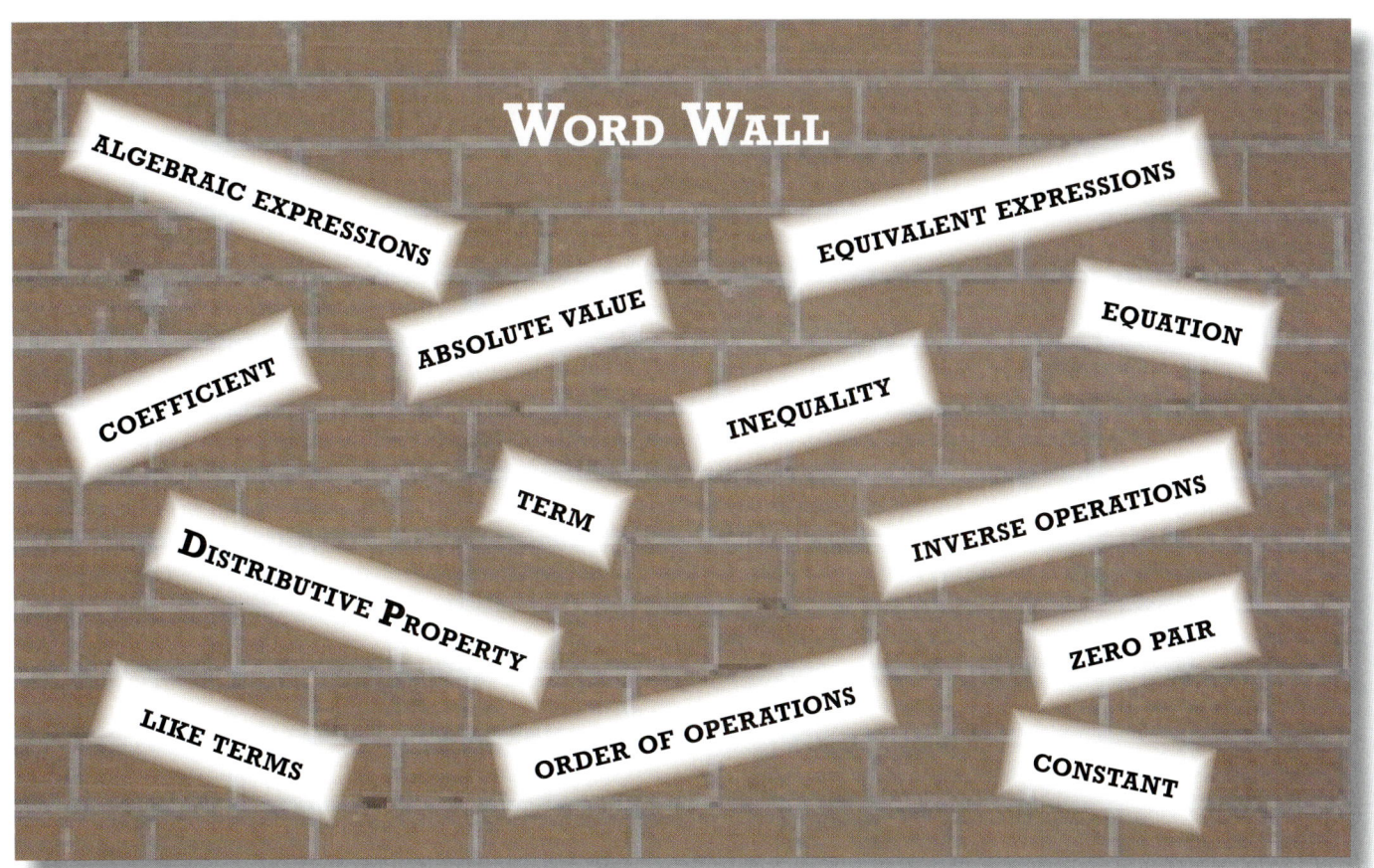

BLOCK 1 ~ EXPRESSIONS AND EQUATIONS
TIC-TAC-TOE

What's the Process?

Create a poster that explains the process of solving a multi-step equation.

See page 24 for details.

Small Business Profits

Study three different business start-up plans. Determine how many items the business will need to sell to break even and make a profit.

See page 33 for details.

Order of Ops Poetry

Write three different poems about the order of operations.

See page 14 for details.

Like Terms Game

Create a matching game requiring players to combine like terms to make pairs.

See page 14 for details.

Math in Everyday Life

Write word problems that can be solved using equations that require two or more steps to solve.

See page 42 for details.

Temperature Systems

Convert temperatures from one system to another.

See page 24 for details.

Equation Mats

Produce a "How To..." guide to show others how to use equation mats to solve a variety of equations.

See page 42 for details.

Compound Inequalities

Solve compound inequalities and graph solutions on a number line.

See page 38 for details.

One Does Not Belong

Make cards with different equations. Ask participants to find the one that does not belong.

See page 38 for details.

ORDER OF OPERATIONS

LESSON 1.1

 Find the value of expressions using the order of operations.

Mr. Marshall asked his students to find the value of the expression
$$7 + 2 \cdot 3 - 1$$
Sasha is positive that the answer is 26. Michelle believes the answer is 12. Mr. Marshall has both students show their work on the board. Look at their work below. Who do you agree with? Why?

Sasha	Michelle
$7 + 2 \cdot 3 - 1$	$7 + 2 \cdot 3 - 1$
$= 9 \cdot 3 - 1$	$= 7 + 6 - 1$
$= 27 - 1$	$= 13 - 1$
$= 26$	$= 12$

Mathematicians have established an **order of operations**. The order of operations is a set of rules which are followed when evaluating an expression with more than one operation. Using the correct order of operations helped Michelle find the correct answer to Mr. Marshall's question because she multiplied before adding or subtracting.

ORDER OF OPERATIONS

1. Find the value of expressions inside grouping symbols such as parentheses, absolute value bars and fraction bars.
2. Find the value of all powers.
3. Multiply and divide from left to right.
4. Add and subtract from left to right.

All operations within grouping symbols must be completed before any of the other steps are completed. The three types of grouping symbols you will work with in this book are parentheses, absolute value bars and fraction bars.

Absolute value is the distance a number is from zero. The absolute value of a number is always positive. For example:
$$|-9 + 4| = |-5| = 5$$

The fraction bar is another symbol used to represent division. Perform all operations in the numerator and denominator before dividing. For example:
$$\frac{2 + 10}{5 - 1} = \frac{12}{4} = 3$$

EXAMPLE 1 Find the value of each expression.
a. $8(-3 + 7) - 2 \cdot 5$
b. $5^2 + |3 - 8| \div 5$

SOLUTIONS

a. Add the integers inside the parentheses. $8(-3 + 7) - 2 \cdot 5 = 8(4) - 2 \cdot 5$
Multiply from left to right. $= 32 - 10$
Subtract. $= 22$

b. Subtract inside the absolute value bars. $5^2 + |3 - 8| \div 5 = 5^2 + |-5| \div 5$
Make the value inside the bars positive. $= 5^2 + 5 \div 5$
Find the value of the power. $= 25 + 5 \div 5$
Divide. $= 25 + 1$
Add. $= 26$

EXAMPLE 2 Find the value of the expression.

$$\frac{(-5 - 3)^2}{6 - 4} - 50$$

SOLUTION

Perform operation inside the parentheses. $\dfrac{(-5 - 3)^2}{6 - 4} - 50 = \dfrac{(-8)^2}{6 - 4} - 50$

Find the value of $(-8)^2$. $= \dfrac{64}{6 - 4} - 50$

The fraction bar is a type of grouping symbol.
Subtract in the denominator before dividing. $= \dfrac{64}{2} - 50$

Divide numerator by denominator. $= 32 - 50$

Subtract. $= -18$

EXAMPLE 3 Jakim's family took a vacation to California. The plane tickets cost a total of $840, the hotel cost $250 and a car rental cost $130. There are 5 people in Jakim's family.
a. Write an expression that could be used to find the cost per person.
b. Find the cost per person.

SOLUTIONS

a. The numbers must be added before dividing by the number of people. Use parentheses or the fraction bar to group the numbers that must be added.

$(840 + 250 + 130) \div 5$
OR
$\dfrac{840 + 250 + 130}{5}$

b. Add the numerators. $\dfrac{840 + 250 + 130}{5} = \dfrac{1220}{5}$

Divide. $\dfrac{1220}{5} = \$244$

The vacation cost $244 per person.

EXERCISES

1. Nathan and Takashi each evaluated the expression $16 - 5 \cdot 2 + 4$. Nathan believes the solution is 10. Takashi disagrees and says the answer is 26.
   a. Who is correct? What operation did he perform first?
   b. What operation was done first by the student who was incorrect?

Evaluate each expression.

2. $18 \div 2 + 7 \cdot 3$

3. $6 \cdot 5 - 4 \div 2$

4. $(3 + 5)^2 - 8 \cdot 3$

5. $45 \div (-3)^2 + 4(2 + 1)$

6. $5 + 4|-2 + 8|$

7. $5(11 + 1) - 3(2 + 11)$

8. $12 - 2 \cdot 10 \div 5 - 1$

9. $|3 - 27| \div |4 + 2|$

10. $21 \div 3 \cdot 7 - 4^2$

Evaluate each expression.

11. $\dfrac{4 + 28}{6 - 2}$

12. $\dfrac{20(7 - 3)}{2 \cdot 5} - 7$

13. $\dfrac{|-18 + 3|}{9 - 6}$

14. $31 - \dfrac{5 + 4 \cdot 5 - 1}{2}$

15. $\dfrac{6(9 - 7)^2}{-2}$

16. $\dfrac{25}{5} - \dfrac{10|5 - 1|}{2}$

17. The Chess Club is selling tickets to Saturday's Winter Ball. Admission with Student ID is $5 per person. Admission without a Student ID is $8. The Chess Club sold 130 tickets to students with an ID Card and 40 tickets to students without ID Cards.
    a. Write an expression to represent the total amount of money the Chess Club collected in ticket sales.
    b. How much money did the Chess Club collect?

18. Four friends ordered Chinese food for dinner. The Kung Pao Chicken cost $11. The Sweet and Sour Pork cost $8 and the Mongolian Beef cost $13. The friends want to split the cost equally. Find the cost per person. Use words and/or numbers to show how you determined your answer.

19. Explain why it is necessary to have an order of operations in mathematics.

20. Create an expression with at least five numbers and two different operations that has a value of 15. Show all work necessary to justify your answer.

Insert one set of parentheses in each numerical expression so that it equals the stated amount. Use mathematics to show that your answer is correct.

21. $6 + 3 + 11 \div 4 = 5$

22. $7 + 1 \cdot 4 - 2 \cdot 5 = 17$

23. $-1 \cdot 6 + 8 - 4 \div 2 + 2 = 1$

EVALUATING EXPRESSIONS

LESSON 1.2

 Use the order of operations to evaluate expressions.

An expression that contains numbers, operations and variables is an **algebraic expression**. Variables are symbols used to represent numbers in algebraic expressions and equations. Algebraic expressions can be evaluated when the values of the variables are given.

> **EVALUATING EXPRESSIONS**
> 1. Rewrite the expression by replacing the variables with the given values.
> 2. Follow the order of operations to compute the value of the expression.

EXAMPLE 1   Evaluate each algebraic expression.
  a. $2x - 8$ when $x = 4$
  b. $\dfrac{-5(m + y)}{m}$ when $m = -2$ and $y = 10$

SOLUTIONS

a. Write the expression. $2x - 8$
   Substitute 4 for x. $2(4) - 8$
   Multiply. $8 - 8$
   Subtract. 0

b. Write the expression. $\dfrac{-5(m + y)}{m}$

   Substitute -2 for m and 10 for y. $\dfrac{-5(-2 + 10)}{-2}$

   Add inside parentheses. $\dfrac{-5(8)}{-2}$

   Multiply. $\dfrac{-40}{-2}$

   Divide the numerator by the denominator. 20

Input, x	$\dfrac{4 + 5x}{2}$	Output
2	$\dfrac{4 + 5(2)}{2}$	7
5	$\dfrac{4 + 5(5)}{2}$	14.5
-6	$\dfrac{4 + 5(-6)}{2}$	-13

Different values can be substituted for the variable in an expression. A table is used to help organize mathematical computations. The values for the variable are often called the input values. The values of the expression are often called the output values.

EXAMPLE 2 Fill in the table by evaluating the given expression for the values listed.

x	$-3x + 7$	Output
-2		
0		
4		
9		

SOLUTION Rewrite the expression with the input value in the place of x and then follow the order of operations. In this case, multiply and then add to find the output values.

x	$-3x + 7$	Output
-2	$-3(-2) + 7$	13
0	$-3(0) + 7$	7
4	$-3(4) + 7$	-5
9	$-3(9) + 7$	-20

An **equation** is a mathematical sentence that contains an equals sign between two expressions. Equations that involve at least one variable are neither true nor false until the equation is evaluated with given values for the variables. Values can be considered the solutions to an equation if they make the equation true.

EXAMPLE 3 State whether each equation is true or false for the values of the variables given.
a. $3x + 2y = 8$ when $x = 2$ and $y = 1$
b. $-5x + 9 = y$ when $x = 6$ and $y = -21$
c. $y = \frac{1}{2}x + 1$ when $x = 8$ and $y = 17$

SOLUTIONS

a. Substitute $x = 2$ and $y = 1$. $3(2) + 2(1) \stackrel{?}{=} 8$
   Multiply. $6 + 2 \stackrel{?}{=} 8$
   Add. $8 = 8$ **TRUE**

b. Substitute $x = 6$ and $y = -21$. $-5(6) + 9 \stackrel{?}{=} -21$
   Multiply. $-30 + 9 \stackrel{?}{=} -21$
   Add. $-21 = -21$ **TRUE**

c. Substitute $x = 8$ and $y = 17$. $17 \stackrel{?}{=} \frac{1}{2}(8) + 1$
   Multiply. $17 \stackrel{?}{=} 4 + 1$
   Add. $17 \neq 5$ **FALSE**

Lesson 1.2 ~ Evaluating Expressions

EXERCISES

Evaluate each expression when $x = 3$.

1. $2x - 1$

2. $\dfrac{-9x + 5}{2}$

3. $\dfrac{1}{3}x - 7$

4. $(x + 3)^2$

5. $5(x - 1)^2 + 2$

6. $-9 - 5x$

Evaluate each expression for the given values of the variables.

7. $\dfrac{3x - 5}{2}$ when $x = 7$

8. $\dfrac{-2(7x - 3)}{5}$ when $x = 4$

9. $8y - 2x$ when $y = 5$ and $x = 5$

10. $\dfrac{3}{4}a + 3$ when $a = 12$

11. $(4 + 3x)^2$ when $x = -1$

12. $-3y - 4m$ when $y = 1$ and $m = \dfrac{1}{2}$

13. $y + \dfrac{x - 7}{5}$ when $y = -2$ and $x = 22$

14. $0.5m - 0.1n$ when $m = 10$ and $n = 10$

Copy each table. Complete each table by evaluating the given expression for the values listed.

15.

x	$6x - 4$	Output
-2		
0		
$\dfrac{1}{2}$		
3		
8		

16.

x	$\dfrac{5x - 3}{2}$	Output
-3		
0		
4		
11		

State whether each equation is true or false for the values of the variables given. Show all work necessary to justify your answer.

17. $5x + 2y = 10$ when $x = 2$ and $y = 0$

18. $-3x + y = 7$ when $x = 1$ and $y = 4$

19. $y = 8x + 9$ when $x = 5$ and $y = 40$

20. $4y = x + 5$ when $x = 7$ and $y = 3$

21. $y = 3(x - 6)$ when $x = 4$ and $y = -6$

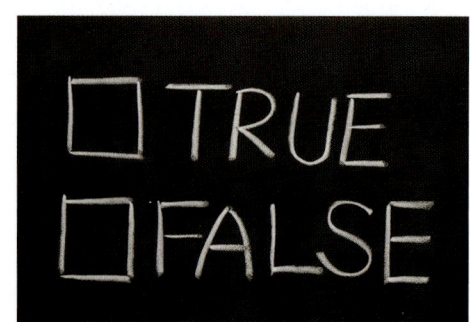

22. $0.5x + 5y = 17.5$ when $x = 5$ and $y = 4$

23. Victor evaluated the expression $5x + 3$ when $x = 4$. His work is shown below. Explain what Victor did wrong and find the correct value of the expression.

$$5x + 3 \rightarrow 54 + 3 = 57$$

24. A teacher described a specific expression to her students that could only have integer input and output values. She wanted the output values to be greater than 9 and less than 13. What are the possible input values, x, if the expression is $\frac{1}{3}x + 4$? Use words and/or numbers to show how you determined your answer.

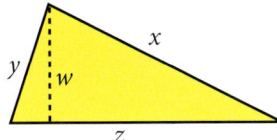
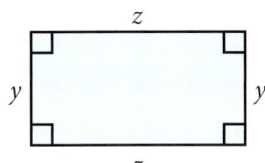
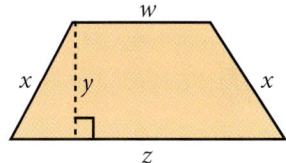

25. Use the formulas given to find the perimeter of each figure above when $w = 2$, $x = 5$, $y = 3$ and $z = 10$.
    a. TRIANGLE   **b.** RECTANGLE   **c.** TRAPEZOID
        $P = x + y + z$ $P = 2(y + z)$ $P = w + 2x + z$

26. Use the formulas given to find the area of each figure above when $w = 2$, $x = 5$, $y = 3$ and $z = 10$.
    a. TRIANGLE   **b.** RECTANGLE   **c.** TRAPEZOID
        $A = \frac{1}{2}wz$ $A = yz$ $A = \frac{1}{2}y(w + z)$

27. Tom went shopping at the mall. He found one type of shirt he liked for $12. He also discovered a pair of shorts for $16. Both the shirt and the shorts came in many different colors.

    a. Let x represent the number of shirts and y represent the number of shorts Tom purchased. Write an algebraic expression that represents the total cost for x shirts and y shorts.
    b. Tom decided to buy three shirts and five pairs of shorts. What was the total cost for his purchase?
    c. Sam, a friend of Tom's, decided to buy the same kind of shorts and shirts. The total cost for his purchase was $80. How many shirts and pairs of shorts might Sam have purchased? Explain how you know your answer is correct.

28. The table at right shows admission prices for Centerville's movie theater.

Centerville Movie Admission	
Adult (18-61 years old)	$8.00
Senior Citizen (62 years and above)	$6.50
Children (5-17 years old)	$4.00
Children (4 years and under)	$2.50

    a. The Johnson family consists of 2 adults, 1 senior citizen and three children (ages 3, 7 and 13). What will be the total cost for admission for the Johnson family to see a movie at the theater?
    b. Jacob is having a birthday bash for his thirteenth birthday. His mom agreed to take Jacob and 9 of his friends to the theater for the party. All of Jacob's friends are also twelve or thirteen. How much will it cost for all the kids, plus Jacob's mom, to go to a movie?
    c. The Smiths spent $26.50 on movie admissions for their family. Give two possible descriptions of the ages of people in the Smith family. Show all work necessary to justify your answer.

REVIEW

Evaluate each expression.

29. $5 + 2 \cdot 7 - 20$

30. $\frac{6 + 9}{3} + 4$

31. $1 - 10 - 11 + 2 \cdot 3^2$

32. $3(2 + 4)^2 - 100$

33. $-5 \cdot 7 + 6(-2 - 1)$

34. $\frac{70 - 10}{6 + 4} - 6$

THE DISTRIBUTIVE PROPERTY

LESSON 1.3

Simplify expressions using the Distributive Property and combining like terms.

Every algebraic expression has at least one term. A **term** is a number or the product of a number and a variable. Terms are separated by addition and subtraction signs. A **constant** is a term that has no variable. The number multiplied by a variable in a term is called the **coefficient**.

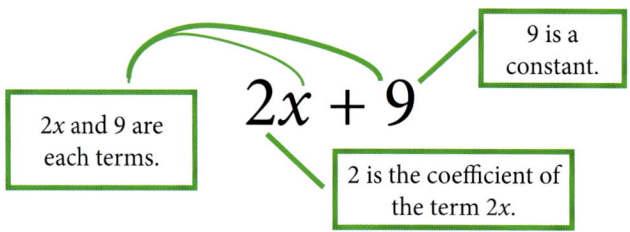

$2x + 9$

- 9 is a constant.
- $2x$ and 9 are each terms.
- 2 is the coefficient of the term $2x$.

Some expressions contain parentheses. One tool that will help you work with these expressions is called the **Distributive Property**. The Distributive Property allows you to rewrite an expression without parentheses. This is done by distributing the front coefficient to each term inside the parentheses. This will be a crucial step in solving equations that contain parentheses.

THE DISTRIBUTIVE PROPERTY

For any numbers a, b and c:
$$a(b + c) = a \cdot b + a \cdot c$$
$$a(b - c) = a \cdot b - a \cdot c$$

EXAMPLE 1   Use the Distributive Property to simplify each expression.
a. $2(x + 6)$ b. $\frac{1}{4}(y - 20)$ c. $-5(3x - 1)$

SOLUTIONS   Drawing arrows from the front coefficient to each term inside the parentheses will help guide you.

a. $2(x + 6)$ $= 2(x) + 2(6)$
              $= 2x + 12$

b. $\frac{1}{4}(y - 20)$ $= \frac{1}{4}(y) - \frac{1}{4}(20)$
                         $= \frac{1}{4}y - 5$

c. $-5(3x - 1)$ $= -5(3x) - (-5)(1)$
                $= -15x + 5$

Watch the signs when distributing a negative number.

The Distributive Property is very useful when doing mental math calculations. Certain numbers are easier to multiply together than others. In **Example 2**, notice how you can rewrite a number as a sum or difference of two other numbers that are easier to work with and then do the math mentally.

EXAMPLE 2 Find each product by using the Distributive Property and mental math.
  a. $4(103)$
  b. $998 \cdot 7$
  c. $8(6.5)$

SOLUTIONS
a. $4(103) = 4(100 + 3)$
   $4(100) + 4(3)$
   $400 + 12 = 412$

b. $998 \cdot 7 = 7(1000 - 2)$
   $7(1000) - 7(2)$
   $7000 - 14 = 6{,}986$

c. $8(6.5) = 8(6 + 0.5)$
   $8(6) + 8(0.5)$
   $48 + 4 = 52$

An algebraic expression or equation may have **like terms** that can be combined. Like terms are terms that have the same variable raised to the same power. The numerical coefficients do not need to be the same. If there are parentheses involved in the expression, the Distributive Property must be used FIRST before combining like terms.

An expression is simplified if it has no parentheses and all like terms have been combined. When combining like terms you must remember that the operation in front of the term (addition or subtraction) must remain attached to the term. Rewrite the expression by grouping like terms together before adding or subtracting the coefficients to simplify.

EXAMPLE 3 Simplify by combining like terms. $3x - 2y + 4 - 2x + x + 4y$

SOLUTION

Mark terms that are alike. $\quad 3x - 2y + 4 - 2x + x + 4y$

Group like terms. $\quad 3x - 2x + x \;\; -2y + 4y \;\; + 4$

Combine. $\quad 2x + 2y + 4$

> A subtraction sign is treated as a negative sign on the coefficient it precedes.

EXAMPLE 4 Simplify by combining like terms. $-5(2x - 1) + 3x - 2$

SOLUTION

Distribute. $\quad -5(2x - 1) + 3x - 2 \;=\; -10x + 5 + 3x - 2$

Group like terms. $\quad = -10x + 3x + 5 - 2$

Combine. $\quad = -7x + 3$

In each example so far in this lesson, the original and simplified expressions are called equivalent expressions. Two or more expressions that represent the same simplified algebraic expression are called **equivalent expressions**.

EXPLORE! MATCH THEM UP

Step 1: On your own paper, copy each expression below. Leave at least two lines between each expression.

- **A.** $2(x + 4) - 5$
- **B.** $3(x - 3) - 2x$
- **C.** $1 + 8(x + 1)$
- **D.** $9 - 4x - 1 - x$
- **E.** $2(4x - 5) + 1 - 7x$
- **F.** $20 - 3(x + 4) - 2x$
- **G.** $3x + 3 + 7x - 8x$
- **H.** $-2(x - 3) + x + 5$
- **I.** $2(x + 7) - 3(x + 1)$
- **J.** $6x + 5x - x - 2x + 9$

Step 2: Simplify each expression.

Step 3: Every expression listed above is equivalent to one other expression in the list. Classify the ten expressions into five groups of equivalent expressions.

Step 4: Create another 'non-simplified' expression for each group that is equivalent to the other expressions in the group.

EXERCISES

Use the Distributive Property to simplify each expression.

1. $5(x + 1)$

2. $\frac{3}{5}(5x + 10)$

3. $2(4m + 5)$

4. $-6(x - 10)$

5. $-3(h - 11)$

6. $16\left(\frac{1}{2}x - 2\right)$

Find the product by using the Distributive Property.

7. $7(105)$

8. $68(10.5)$

9. $896 \cdot 4$

10. Melissa said terms could be combined in the expression shown below becasue they have the same coefficients. Do you agree or disagree? Explain your reasoning.

$$5y + 6b + 5 + 6p$$

11. Tasha finds 7 DVDs she wants to purchase at the video store. Each DVD is $14.95.
   a. Show how Tasha could use the Distributive Property to help mentally calculate the total cost of the DVDs.
   b. How much will she pay for the seven DVDs?

Simplify each expression.

12. $9 + 2x - 4 + 8x$

13. $7(x - 2) + 6(x + 1)$

14. $7 + 3(x - 4)$

15. $-9x + 8x + 7y - 6y$

16. $7x + 3y - x + 4y - 2x$

17. $12y - 3x + 10x - y$

18. $-10(3x + 2) - 12$

19. $9x - 3(x + 2) + 7$

Write and simplify an expression for the perimeter of each figure.

20.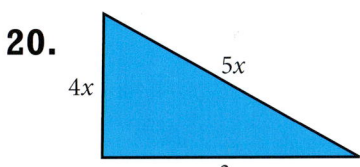

21.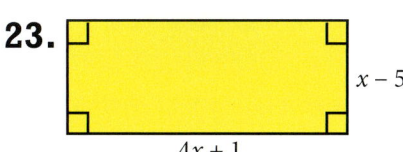

22.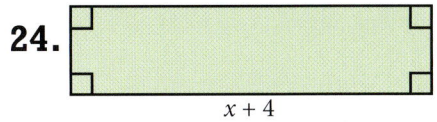

23.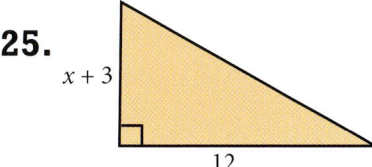

Write and simplify an expression for the area of each figure. Show all work necessary to justify your answer.

Area of Rectangle = *length* · *width*   Area of Triangle = $\frac{1}{2}$ · *base* · *height*

24.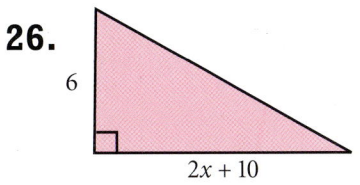

25.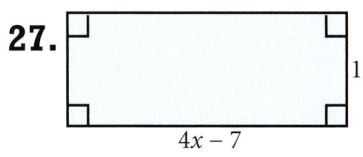

26.

27.

In each set of three expressions, two are equivalent. Find the equivalent expressions. Show all work necessary to justify your answer.

28. A. $3x + 4 - 2 + 7x$
    B. $18x - 5 - 8x + 7$
    C. $2x + 8x - 4 + 2$

29. A. $3(x - 1) + 12$
    B. $4(x + 3) - 1$
    C. $4(x + 2) - x + 1$

30. Whitney states that $8x - 7$ is equivalent to $7 - 8x$. Do you agree with her? Explain your answer.

Lesson 1.3 ~ The Distributive Property   13

REVIEW

31. Evaluate each expression when $x = 2$.

  a. $5(x + 8)$ **b.** $\dfrac{3x - 2}{6}$

  c. $(x + 5)^2 + 2x$ **d.** $16\left(\dfrac{1}{2}x + 1\right)$

32. Frances earns $20 per day for yard work plus $4 more for every pound of yard debris she removes.

  a. Write an expression that represents the total amount Frances would be paid for a day of yard work when she disposed of x pounds of yard debris.

  b. How much would Frances get paid on a day when she removed 18 pounds of yard debris? Use words and/or numbers to show how you determined your answer.

Tic-Tac-Toe ~ Order of Ops Poetry

Poetry is an art form that is composed of carefully chosen words to express a greater depth of meaning. Poetry can be written about many different subjects, including mathematics. The order of operations is a key element in mathematics so that all students, teachers and mathematicians reach the same answer when calculating the value of the same expression. Write two different poems about the order of operations. One should be an acrostic poem and the other should be a quatrain. Research and find another style of poetry. Write one more poem about the order of operations using this style.

Quatrain
A poem consisting of four lines. Lines 2 and 4 must rhyme and have a similar number of syllables.

Acrostic
A poem where certain letters, usually the first in each line, form a word or message when read in a sequence.

Tic-Tac-Toe ~ Like Terms Game

Write 15 variable expressions containing like terms that have not been combined. Make sure you have a minimum of five expressions containing parentheses. Cut thicker paper (such as cardstock, construction paper, index cards or poster board) into 30 equal-sized pieces. Write each expression on one card. On another card, write the expression in simplest form. Use these cards to play a memory game with a friend, classmate or family member. Record each pair of cards that each participant wins on a sheet of paper by listing the non-simplified and simplified expressions. Turn in the cards and the game sheet to your teacher.

SOLVING ONE-STEP EQUATIONS

LESSON 1.4

 Use inverse operations to solve one-step equations.

In order to solve a mathematical equation, the variable must be isolated on one side of the equation and have a front coefficient of one. This process is sometimes referred to as "getting the variable by itself". The most important thing to remember is that the equation must always remain balanced. Whatever occurs on one side of the equals sign MUST occur on the other side so the equation remains balanced. The properties of equality are listed below. Note that you can perform any of the four basic operations to an equation as long as that operation is done to both sides of the equation.

THE PROPERTIES OF EQUALITY

For any numbers a, b and c:
If $a = b$, then $a + c = b + c$ (Addition Property of Equality)
If $a = b$, then $a - c = b - c$ (Subtraction Property of Equality)
If $a = b$, then $a \cdot c = b \cdot c$ (Multiplication Property of Equality)
If $a = b$, then $\frac{a}{c} = \frac{b}{c}$ (Division Property of Equality)

EXPLORE!   **ADDITION AND SUBTRACTION EQUATIONS**

Each blue chip represents the integer +1. Each red chip represents the integer −1. When a positive integer chip is combined with a negative integer chip, the result is zero. This pair of integer chips is called a **zero pair**.

Step 1: On your equation mat, place the variable cube on one side with 4 positive integer chips. On the other side of the mat, place 9 positive integer chips. Write the equation that is represented by the items on the mat.

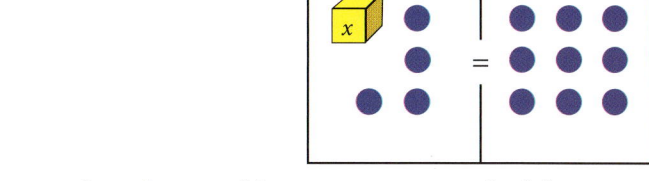

Step 2: In order to isolate the variable, you must get rid of the 4 integer chips with the variable. If you take chips off one side of the equation, you must do the same on the other side. How many chips are left on the right side? What does this represent?

Step 3: Clear your mat and place chips on the mat to represent the equation $x - 2 = 6$. Draw this on a sheet of paper.

Step 4: How could you "get rid of" the chips that are with the variable? How many chips end up on the opposite side of the mat when you cancel out the 2 negative integer chips that are with the variable? Write your answer in the form $x = \_\_\_$.

EXPLORE! (CONTINUED)

Step 5: Clear your mat and place chips on the mat to represent the equation $-3 + x = -1$. Draw this on your own paper.

Step 6: Isolate the variable by canceling out chips on the variable side of the equation. Remember that whatever you do to one side of the equation you must do to the other side. What does x equal in this case?

Step 7: Create your own equation on your mat. Record the algebraic equation on your paper.

Step 8: Solve your equation. What does your variable equal?

Step 9: Write a few sentences that describe how to solve a one-step addition or subtraction equation using equation mats, integer counters and variable cubes.

You will not always have integer chips or an equation mat available to use when you are solving equations. You can solve equations using the balancing method on paper. When isolating the variable, remember to perform an **inverse operation** on both sides of the equation. Inverse operations are operations that undo each other, such as addition and subtraction.

EXAMPLE 1

Solve for x. Check the solution.

a. $x - 6 = 22$ b. $8x = 88$

c. $x + 10 = -7$ d. $\dfrac{x}{5} = 21$

SOLUTIONS

Draw a vertical line through the equals sign to help you stay organized. Whatever is done on one side of the line to cancel out a value must be done on the other side.

a. $x - 6 = 22$
   $+6+6$
   $\boxed{x = 28}$

Check the answer by substituting the answer back into the equation.

✓ $28 - 6 \stackrel{?}{=} 22$
   $22 = 22$

b. $\dfrac{8x}{8} = \dfrac{88}{8}$
   $\boxed{x = 11}$

✓ $8(11) \stackrel{?}{=} 88$
   $88 = 88$

c. $x + 10 = -7$
   $-10-10$
   $\boxed{x = -17}$

✓ $-17 + 10 \stackrel{?}{=} -7$
   $-7 = -7$

d. $5 \cdot \dfrac{x}{5} = 21 \cdot 5$
   $\boxed{x = 105}$

✓ $\dfrac{105}{5} \stackrel{?}{=} 21$
   $21 = 21$

Equations in this lesson have been written symbolically. There will be times when an equation will be written in words and need to be translated into mathematical symbols in order to solve the equation. Here are some key words you need to remember when translating words into math symbols:

Addition	Subtraction	Multiplication	Division	Equals
sum increased by more than plus	difference decreased by less than minus	product multiplied by times of	quotient divided by	is equal to is

EXAMPLE 2 Write an equation for each statement. Solve each equation and check your solution.
a. The product of eleven and a number is seventy-seven.
b. The sum of a number and 52 is 98.
c. A number decreased by 13 is 214.

SOLUTIONS

a. Write the equation. "Product" means multiplication.

$$11x = 77$$
$$\frac{11x}{11} = \frac{77}{11}$$

Divide both sides of the equation by 11.
The number is 7.

$$x = 7$$

☑ Check the answer. $11 \cdot 7 = 77$

b. Write the equation. "Sum" means addition.

$$p + 52 = 98$$
$$\begin{array}{r} p + 52 = 98 \\ -52 \quad -52 \end{array}$$

Subtract 52 from both sides of the equation.
The number is 46.

$$p = 46$$

☑ Check the answer. $46 + 52 = 98$

c. Write the equation. "Decreased by" means subtraction.

$$y - 13 = 214$$
$$\begin{array}{r} y - 13 = 214 \\ +13 \quad +13 \end{array}$$

Add 13 to both sides of the equation.
The number is 227.

$$y = 227$$

☑ Check the answer. $227 - 13 = 214$

EXAMPLE 3 The sum of Kirk's age and his dad's age is 53. Kirk is 14. Write an equation that represents this situation using d to represent the dad's age. Solve the equation and check the solution.

SOLUTION

Write the equation.
Subtract 14 from both sides of the equation.

$$\begin{array}{r} 14 + d = 53 \\ -14 \quad -14 \end{array}$$
$$d = 39$$

☑ Check the answer. $14 + 39 = 53$

Kirk's dad is 39 years old.

EXERCISES

Solve each equation for *x*. Show all work necessary to justify your answer.

1. $x - 4 = 27$
2. $-4x = 84$
3. $6x = 54$
4. $-12 + x = -30$
5. $96 = 35 + x$
6. $\dfrac{x}{5} = 8$
7. $\dfrac{x}{6} = -12$
8. $x - 3 = -9$
9. $-110 = -5x$

10. Shawn did not check his solutions for the four-problem quiz on one-step equations. Check Shawn's answers. If the answer is incorrect, find the correct answer.

    a. $x + 82 = 124$
       Shawn's answer: $x = 206$

    b. $-7 + x = 12$
       Shawn's answer: $x = 19$

    c. $9x = 63$
       Shawn's answer: $x = -7$

    d. $\dfrac{x}{4} = 8$
       Shawn's answer: $x = 2$

11. Patty wants to use an equation mat, integer chips and variable cubes to illustrate a one-step multiplication equation. Draw a picture of how she could illustrate the equation $3x = 12$ on the mat. Illustrate how the integer chips can be separated into three equal parts on the mat to show the solution to the equation. State the value of *x*.

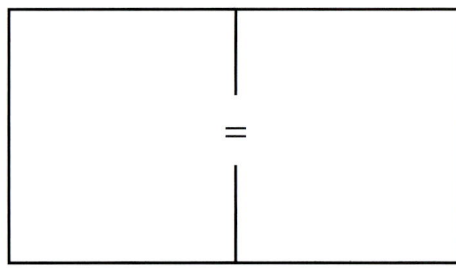
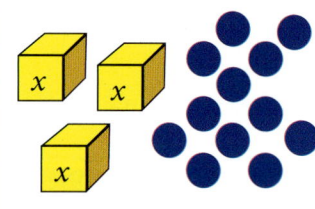

Write an equation for each statement and solve for the variable. Show all work necessary to justify your answer.

12. The product of 8 and a number is 48.

13. The sum of a number and -53 is 89.

14. A number divided by -7 is -7.

15. Sixteen less than a number is 102.

16. A number increased by $1\tfrac{1}{2}$ is $2\tfrac{3}{4}$.

17. One-third of a number is 6.

18. Mike is thinking of two numbers. Their difference is 12. If one of the numbers is 19, what is the other number? Is there only one answer? If not, what is the other possibility for the second number? Explain your reasoning.

19. It is a common belief that one human year is equal to 7 dog-years.
    a. Write an expression that will calculate how old a dog is in dog-years based on its normal "human year" age. Use x to represent the number of human years.
    b. If a dog is 84 dog-years old when he passes away, how old was he in human years? Use words and/or numbers to show how you determined your answer.

Solve each equation for x. Show all work necessary to justify your answer.

20. $x + 1.73 = 2.81$
21. $-0.1x = 16$
22. $\frac{2}{3}x = 10$
23. $\frac{6}{11}x = \frac{21}{22}$
24. $-2.5 + x = 6.75$
25. $-x = 3.4$

26. Julie weighs $5\frac{1}{2}$ times as much as her baby brother, Jeremie. Julie weighs $60\frac{1}{2}$ pounds. How much does Jeremie weigh? Show all work necessary to justify your answer.

27. Harley argues that he can always use mental math to solve one-step equations so he should not have to show his work. What might be a good argument that he should show his work?

REVIEW

Evaluate each expression.

28. $12 + 12 \div 3 \cdot 5 - 1$
29. $\frac{(3+7)^2}{5} + 6$
30. $3(-21 + 10) + 7$
31. $-4\left(\frac{1}{2} + \frac{1}{2}\right) - 2$

Use the Distributive Property to simplify each expression.

32. $6(x - 3)$
33. $-2(9x - 1)$
34. $-4(5m + 7)$
35. $\frac{1}{4}(8x - 4)$

36. Evaluate the following using mental math.

    a. $-3 + -2$
    b. $-4(6)$
    c. $-5 + 11$
    d. $6 - 9$
    e. $27 \div 3$
    f. $(-20)(-5)$
    g. $-8 + 1$
    h. $\frac{-36}{6}$

SOLVING TWO-STEP EQUATIONS

LESSON 1.5

 Solve two-step equations.

Jim took a summer road trip across the country. He started his trip 20 miles east of Seattle. He headed east on I-90. Every hour (h) he traveled 60 miles further away from Seattle. The expression that represents his distance from Seattle is:
$$d = 20 + 60h$$

Jim forgot to bring a watch, but noticed that he was 500 miles from Seattle after his first day of driving. Assuming Jim traveled at an equal rate, how many hours did Jim drive on the first day of his road trip?

To solve this problem, you must be able to solve a two-step equation. Since Jim has driven 500 miles, substitute that value for d in the equation:
$$500 = 20 + 60h$$

To isolate the variable in a two-step equation you must perform two inverse operations. The inverse operations must undo the order of operations. That means you start by undoing addition or subtraction. Then use inverse operations to remove any multiplication or division.

You can determine how long Jim has been driving by solving the equation $500 = 20 + 60h$.

Write an equation.
Subtract 20 from both sides of the equation.
Divide both sides of the equation by 60.

$$\begin{array}{r|l} 500 = \cancel{20} + 60h \\ -20 & -\cancel{20} \\ \hline \dfrac{480}{60} & \dfrac{\cancel{60}h}{\cancel{60}} \end{array}$$

$$\boxed{8 = h}$$

Jim drove 8 hours on his first day.

EXAMPLE 1 Solve the equation for x. Check the solution. $8x - 3 = 85$

SOLUTION Add 3 to both sides of the equation.

Divide both sides of the equation by 8.

$$\begin{array}{r|l} 8x - \cancel{3} = 85 \\ +\cancel{3} & +3 \\ \hline \dfrac{\cancel{8}x}{\cancel{8}} & \dfrac{88}{8} \end{array}$$

$$\boxed{x = 11}$$

☑ Check the answer.

$$8(11) - 3 \stackrel{?}{=} 85$$
$$88 - 3 \stackrel{?}{=} 85$$
$$85 = 85$$

EXAMPLE 2 Solve the equation for *x*. Check the solution. $-2 = \frac{x}{9} + 5$

SOLUTION

Subtract 5 from both sides of the equation.
$$\begin{array}{r} -2 = \frac{x}{9} + 5 \\ -5 \quad\quad -5 \\ \hline \end{array}$$

The variable can be on either side of the equals sign.

Multiply both sides of the equation by 9. $9 \cdot (-7) = \frac{x}{9} \cdot 9$

$$\boxed{-63 = x}$$

☑ Check the solution.
$$-2 \stackrel{?}{=} \frac{-63}{9} + 5$$
$$-2 \stackrel{?}{=} -7 + 5$$
$$-2 = -2$$

EXAMPLE 3 Use an equation mat to illustrate and solve the equation $2x - 4 = 6$.

SOLUTION Lay out the variable cubes and integer chips to match the equation.

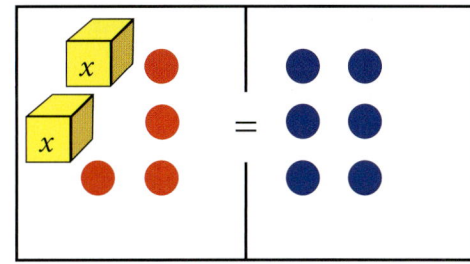

Remove the integer chips from the side with the variable by canceling out four negative integer chips with four positive integer chips (add 4 positive chips to each side).

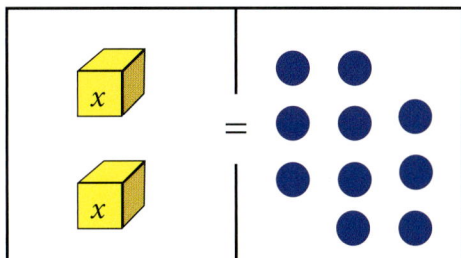

Divide the integer chips on the right side of the mat equally between the two cubes.

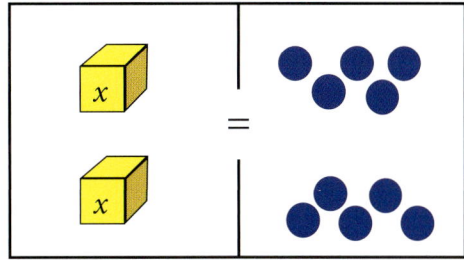

Write the solution. $x = 5$

EXAMPLE 4   Eight more than 3 times a number is 29. Write and solve an equation to find the value of the number.

SOLUTION   Write the equation.
Three times a number $3x$
Eight more than $3x + 8$
Is 29 $3x + 8 = 29$

The word 'is' often means = in math.

$$3x + 8 = 29$$
$$-8 -8$$
$$\frac{3x}{3} = \frac{21}{3}$$
$$\boxed{x = 7}$$

☑ Check the solution.
$$3(7) + 8 \stackrel{?}{=} 29$$
$$21 + 8 \stackrel{?}{=} 29$$
$$29 = 29$$

EXERCISES

Solve each equation for *x*. Show all work necessary to justify your answer.

1. $4x + 2 = 18$

2. $\frac{x}{3} - 4 = 26$

3. $\frac{x}{5} - 7 = 3$

4. $102 = 20x - 8$

5. $-8 + 10x = 102$

6. $\frac{1}{2}x + \frac{3}{4} = 1$

7. $-18 = \frac{x}{3} + 2$

8. $2.5x + 7.5 = 20$

9. $32 = -8 + 4x$

10. The Fahrenheit and Celsius scales are related by the equation: $F = \frac{9}{5}C + 32$

   a. The lowest temperature in Alaska, $-62°$ C, was recorded on January 23, 1971 at Prospect Creek Camp. Use the formula to convert the record temperature to Fahrenheit.

   b. In February, the average high temperature in Puerto Vallarta, Mexico is 81° F. What is this temperature in degrees Celsius? Use mathematics to justify your answer.

11. Jordin solved three problems incorrectly. Describe the error she made in each problem; then find the correct answers.

a.
$$3x - 6 = 27$$
$$-6 -6$$
$$\frac{3x}{3} = \frac{21}{3}$$
$$x = 7$$

b.
$$45 = 10x + 15$$
$$-15 -15$$
$$\frac{45}{-5} = \frac{-5x}{-5}$$
$$x = -9$$

c.
$$\frac{x}{8} - 2 = 22$$
$$+ 2 + 2$$
$$8 \div \frac{x}{8} = 24 \div 8$$
$$x = 3$$

22   Lesson 1.5 ~ Solving Two-Step Equations

Write an equation for each statement. Solve the equation. Show all work necessary to justify your answer.

12. Seven more than twice a number is 25.

13. Six less than the quotient of a number and 3 is −1.

14. Twelve decreased by 5 times a number is 72.

15. Four more than one-half a number is 3.

16. Barry begins the year with $25 in his piggy bank. At the end of each month, Barry adds $3.
   a. How much will Barry have after 5 months have passed? Use words and/or numbers to show how you determined your answer.
   b. Write a formula that could be used to calculate Barry's total savings (S) based on how many months (m) he has deposited money in his bank.
   c. Use your formula to determine how many months have passed when Barry reaches $100 in his account.

17. Mariah's parents got into a car accident. They took their car to the shop to be repaired. When the car was finished, they received a bill of $637. The total cost for parts was $280 and the cost of labor was $42 per hour. Determine how many hours the mechanics spent working on the car. Explain in words how you found your answer.

18. Ryan sells cars for a living. He gets paid $50 per day plus a commission of 2% of the total cost of each car he sells. Ryan's daily earnings (E) can be represented by the equation $E = 0.02x + 50$ where x is the amount of his sales for the day.
   a. What does the 0.02 represent in the equation?
   b. How much did Ryan make on Monday if he only sold one car for $4,400?
   c. Ryan wants to make $1,000 in a day. What must his daily sales be to reach this goal? Show all work necessary to justify your answer.

19. Gemma has three less than two-thirds the number of fish that Polly owns. Gemma has 15 fish. How many does Polly own? Show all work necessary to justify your answer.

20. Which of the following equations has a solution of $x = 6$? Select all that apply.

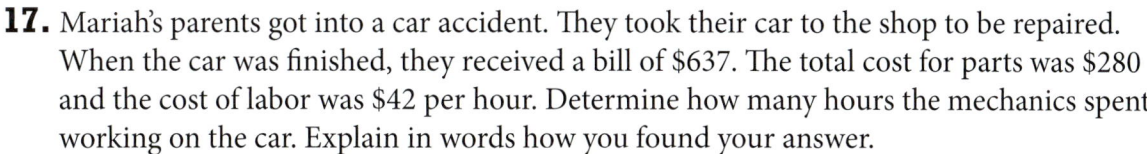

$3x - 5 = 13$ $\frac{1}{3}x - 1 = 2$ $12 = \frac{x}{4} + 11.5$ $-1 - 4x = -25$

21. Nigel spent $3.75 on one bottle of tea and 3 packages of sunflower seeds. The sunflower seeds cost $0.89 per package. How much did the bottle of tea cost? Explain how you know your answer is correct.

22. A delivery service charges a flat rate of $4.25 per box plus an additional amount *per ounce*. Carl mailed a four-pound package at a cost of $10.65. How much did Carl have to pay per ounce? Use words and/or numbers to show how you determined your answer.

REVIEW

Copy each line and insert any combination of the four operations (+, −, ×, ÷) to make each statement true. Use mathematics to show that your answer is correct.

23. 2   3   5 = 17

24. 12   6   10 = −8

25. −1   2   2 = 0

26. 20   5   50   5 = 45

Simplify each expression.

27. $9(x − 2)$

28. $−7 + 6(x − 5) + 2x$

29. $2x + 4x + 7x − 3x$

30. $2(x − 1) + 3(x + 1)$

31. $4(x + 8) − 12$

32. $−5(x + 3) + 5x + 13$

TIC-TAC-TOE ~ TEMPERATURE SYSTEMS

There are three common units of temperatures: Fahrenheit, Kelvin and Celsius.

1. Research the three common units of temperature. Where was each one invented? Where is each one most often used? How are the units related to each other? The following relationships are used to convert one unit of temperature to another:

$F = 1.8C + 32$ $K = C + 273.15$

2. Convert the following units from one system to another.
   a. 20° C to Fahrenheit   **b.** 20° C to Kelvin
   c. 78° F to Celsius   **d.** 289.5 K to Celsius
   e. −4° F to Celsius   **f.** 315 K to Fahrenheit

3. Develop an equation that will convert Fahrenheit to Kelvin.

TIC-TAC-TOE ~ WHAT'S THE PROCESS?

Solving a multi-step equation can be a complicated task depending on the difficulty of the equation. Create a poster to help classmates through the process step-by-step. Include at least two examples on your poster. Address what a student should do if the following items show up in their equation:

- Multiplication or Division
- Addition or Subtraction
- Variables on Different Sides of the Equals Sign
- Variables on the Same Side of the Equals Sign
- Parentheses

SOLVING MULTI-STEP EQUATIONS

LESSON 1.6

 Simplify and solve multi-step equations.

Two different stores at the beach rent bicycles. One store charges an initial fee of $4 plus $2 per hour. This can be represented by the expression $4 + 2h$ where h is the number of hours the bike is rented. The other store charges an initial fee of $10 but only charges $0.50 per hour. This situation is represented by the expression $10 + 0.5h$ where h is the number of hours the bike is rented. At what point would the two bike rentals cost the exact same amount?

The times would be the same when the two expressions are equal.
$$4 + 2h = 10 + 0.5h$$

To solve this equation, you must first get the variables on the same side of the equation. To do this, move one variable term to the opposite side of the equation using inverse operations. It is easiest to move the variable term with the smaller coefficient. This often helps you deal with fewer negative numbers. After the variables are on the same side of the equation, solve the two-step equation that remains.

Subtract $0.5h$ from both sides of the equation.

Subtract 4 from both sides of the equation.
Divide both sides of the equation by 1.5.

$$\begin{array}{r|r} 4 + 2h = 10 + 0.5h \\ -0.5h & -0.5h \\ \hline 4 + 1.5h = 10 \\ -4 & -4 \\ \hline \dfrac{1.5h}{1.5} & \dfrac{6}{1.5} \end{array}$$

$$h = 4$$

The two bike rentals cost the same amount at 4 hours.

To check the solution, substitute the value into each side of the equation. If both sides are equal, the solution is correct.

First store for 4 hours: ☑ $4 + 2(4) = \$12$
Second store for 4 hours: ☑ $10 + 0.5(4) = \$12$

SOLVING MULTI-STEP EQUATIONS

1. Simplify each side of the equation by distributing and combining like terms, when necessary.
2. If variables are on both sides of the equation, balance the equation by moving one variable term to the opposite side of the equals sign using inverse operations.
3. Follow the process for solving one- and two-step equations to get the variable by itself.

EXPLORE!   **MULTI-STEP EQUATIONS**

Step 1: Write the equation that is represented by the equation mat shown below.

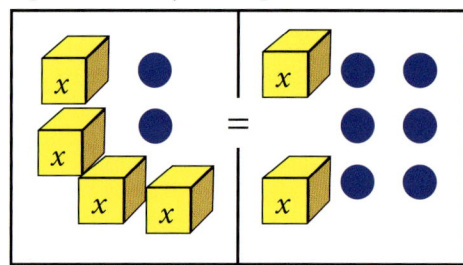

Step 2: There are variable cubes on both sides of the mat. Remove the same number of cubes from each side so that only one side has variable cubes remaining. Draw a picture of what is on the mat now.

Step 3: What is the next step you must take to balance the equation mat? Remember that your goal is to isolate the variable. Draw a picture of what is on the mat now.

Step 4: Once the variables are isolated, x can be determined by dividing the integer chips on the opposite side equally between the remaining variable cubes. How many integer chips belong to each variable cube? What does this represent?

Step 5: Draw a representation of $5x + 7 = 2x + 4$ on an equation mat.

Step 6: Repeat **Steps 2 - 4**. What is the solution to this equation?

EXAMPLE 1   Solve the equation for x. $4(2x - 7) = 20$

SOLUTION

Distribute.

Add 28 to both sides of the equation.

Divide both sides of the equation by 8.

$$4(2x - 7) = 20$$
$$8x - 28 = 20$$
$$+ 28 + 28$$
$$\frac{8x}{8} = \frac{48}{8}$$

$$\boxed{x = 6}$$

☑ Check the solution.

$$4(2 \cdot 6 - 7) \stackrel{?}{=} 20$$
$$4(12 - 7) \stackrel{?}{=} 20$$
$$4(5) \stackrel{?}{=} 20$$
$$20 = 20$$

EXAMPLE 2   Solve the equation for *x*. $-2x + 9 = 4x - 15$

SOLUTION

There are no parentheses so there is no need to use the Distributive Property.

Add 2*x* to both sides of the equation.

Add 15 to both sides of the equation.

$$\begin{array}{r} -2x + 9 = 4x - 15 \\ +2x +2x \\ \hline 9 = 6x - 15 \\ +15 +15 \\ \hline \dfrac{24}{6} = \dfrac{6x}{6} \end{array}$$

Divide both sides of the equation by 6.

$\boxed{4 = x}$

☑ Check the solution.

$$-2(4) + 9 \stackrel{?}{=} 4(4) - 15$$
$$-8 + 9 \stackrel{?}{=} 16 - 15$$
$$1 = 1$$

EXAMPLE 3   Solve the equation for *x*. $5(x + 8) = -2(x - 13)$

SOLUTION

Use the Distributive Property to remove the parentheses.

Add 2*x* to both sides of the equation.

$$\begin{array}{r} 5(x+8) = -2(x-13) \\ 5x + 40 = -2x + 26 \\ +2x +2x \\ \hline 7x + 40 = 26 \\ -40 -40 \\ \hline \dfrac{7x}{7} = \dfrac{-14}{7} \end{array}$$

Subtract 40 from both sides of the equation.

Divide both sides of the equation by 7.

$\boxed{x = -2}$

☑ Check the solution.

$$5(-2 + 8) \stackrel{?}{=} -2(-2 - 13)$$
$$5(6) \stackrel{?}{=} -2(-15)$$
$$30 = 30$$

EXAMPLE 4

Katie opened a coffee cart to earn some extra money. Her one-time equipment start-up cost was $460. It costs her $1 to make each cup of coffee. She plans to sell the cups of coffee for $3. How many cups will she need to sell before she breaks even?

SOLUTION

Let *x* represent the number of cups of coffee sold.

Write an equation that represents the situation.
$$460 + 1x = 3x$$

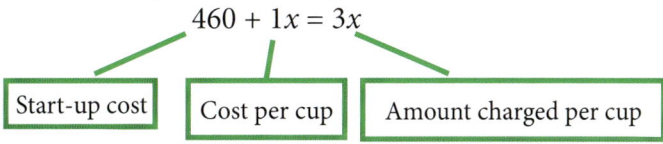

EXAMPLE 4 SOLUTION (CONTINUED)

Subtract $1x$ from both sides of the equation.

Divide both sides by 2.

$$460 + 1x = 3x$$
$$\underline{-1x \quad -1x}$$
$$\frac{460}{2} = \frac{2x}{2}$$

$$230 = x$$

☑ Check the solution.

$$460 + 1(230) \stackrel{?}{=} 3(230)$$
$$460 + 230 \stackrel{?}{=} 690$$
$$690 = 690$$

Katie must sell 230 cups of coffee before she will break even.

EXERCISES

1. Draw an equation mat with integer chips and variable cubes that represents the equation $x + 9 = 3x + 1$. Describe in words or pictures how you would move items around on the mat to solve this equation. What is the solution to this equation?

Solve each equation. Show all work necessary to justify your answer.

2. $2(3x + 6) = 42$

3. $10x = 4x + 66$

4. $-3(x + 4) = -15$

5. $9x + 20 = 34 - 5x$

6. $5x + 3 - 2x = 3$

7. $\frac{1}{2}(6x - 8) = 41$

8. Explain the steps to finding the solution for a multi-step equation.

9. Kelsey made cupcakes to sell as a fundraiser for her trip to Washington DC. She purchased one cupcake pan for $12. She calculated that the ingredients cost her $0.25 per cupcake. She plans to sell the cupcakes for $0.75 each.
  a. Write an equation that could be used to find the number of cupcakes (x) she will need to sell to break even.
  b. Solve the equation from **part a**. Explain how you know your answer is correct.
  c. If she sold 184 cupcakes, what was her profit?

Solve each equation. Show all work necessary to justify your answer.

10. $7x - 9 = -4x + 90$

11. $\frac{1}{4}x + 12 = \frac{1}{2}x + 10$

12. $3(x - 6) = 6x - 90$

13. $2x + 7x + 10 = 11x$

14. $6(x + 2) = 2(2x + 4)$

15. $4.6x - 8.5 = 1.3x + 1.4$

16. $5x = 10x - 4x + 7$

17. $3(2x - 5) = 2x + 1$

18. $x - \frac{1}{3} = 3x + \frac{2}{3}$

19. Ramona has $800 in her account and plans to take $40 out each week. Xavier has $310 in his account and plans to add $30 each week. After how many weeks will Ramona and Xavier have the same amount in their accounts? Use numbers, symbols and/or words to show how you determined your answer.

20. Ty has two job offers at a gym. One has a base monthly salary of $1200 and an additional $50 per new client he gets to join the gym. The other offer does not have a base salary but he gets $100 for every new client who joins the gym. How many clients would Ty need to get to join the gym to make the second offer better than the first? Explain how you know your answer is correct.

21. Francisco made only one mistake on his homework. Describe the mistake he made and solve the equation correctly.

22. An internet movie rental company has two different options for renting movies.
   a. Copy the table below and fill in the total amount paid for movies rented under each plan.

> Option A: Pay a one-time $20 membership fee and then pay $0.50 for each movie rental.
> Option B: Do not become a member and pay $3 for each movie rental.

Movies Rented	Total Cost Option A	Total Cost Option B
0	$20	$0
1		
2		
3		
4		
5		
6		
7		
8		
9		

   b. Write an expression to represent the total cost for **Option A**. Use x for the number of videos rented.
   c. Write an expression to represent the total cost for **Option B**. Use x for the number of videos rented.
   d. Set the two expressions equal to each other. Solve for x.
   e. Explain what the answer to **part d** means in real life. Does the table in **part a** support your answer?
   f. Sal thinks he will rent about 15 movies this year. Which plan would be a better deal for him? Explain your reasoning.

REVIEW

State whether each equation is true or false for the values of the variables given.

23. $5x - 2y = 10$ when $x = 2$ and $y = 0$

24. $4x + 1 = y$ when $x = 5$ and $y = 46$

25. $2(x + 3) = y$ when $x = 3$ and $y = 9$

26. $y = -4(x + 10)$ when $x = -6$ and $y = -16$

SOLUTIONS TO LINEAR EQUATIONS

LESSON 1.7

 Determine if a linear equation in one variable has no solution, one solution or infinitely many solutions.

EXPLORE! **WHAT WORKS?**

Linear equations can have one solution, no solution or infinitely many solutions (meaning any number would make the equation true).

$$2(x + 5) = 2x + 10 \qquad 5x - 11 = 3x - 3$$

$$14 + 3x - 5 = 2x + 3 + x$$

Step 1: Choose an equation out of the green box above. Solve the equation for x. What do you notice about the solution? Do you think your equation has one solution, no solution or infinitely many solutions? Why?

Step 2: Find a partner in the room who solved the same equation as you did in **Step 1**. Do you agree on the number of solutions for your equation? If not, determine whose reasoning is better.

Step 3: With your partner, solve another equation from the box. How many solutions do you think this equation has based on your work? Why?

Step 4: Solve the last equation in the box. There is one of each type (one, no or infinitely many solutions). Based on your work, decide which equation is each type.

Step 5: How can you tell how many solutions a linear equation has based on your work solving the equation? Write two or three summary statements about your findings.

EXAMPLE 1 Solve $6x + 12 = 2x - 5 + 4x + 17$ for x. Describe the number of solutions.

SOLUTION

Group like terms and simplify.

$$6x + 12 \stackrel{?}{=} (2x + 4x) + (-5 + 17)$$
$$6x + 12 \stackrel{?}{=} 6x + 12$$

Move variables to the same side of the equation.

$$\underline{-6x \qquad\quad -6x}$$
$$12 \stackrel{?}{=} 12$$

The ending statement is true for all values of x, therefore the equation has **infinitely many solutions**.

EXAMPLE 2 Solve $3x + 4 + 7x = 5(2x + 1)$ for x. Describe the number of solutions.

SOLUTION

Group like terms.
Distribute and simplify.
Move variables to the same side
of the equation.

$$3x + 7x + 4 \stackrel{?}{=} 5(2x + 1)$$
$$\cancel{10x} + 4 \stackrel{?}{=} \cancel{10x} + 5$$
$$\underline{-\cancel{10x} \qquad -\cancel{10x}}$$
$$4 \stackrel{?}{=} 5$$

Four does not equal five so there are no values of x that make this equation true.

The ending statement is false for all values of x, therefore the equation has **no solution**.

EXAMPLE 3 Solve $8(x + 2) - 5 = 6(x - 1) + 23$ for x. Describe the number of solutions.

SOLUTION

Distribute.

$$8(x + 2) - 5 \stackrel{?}{=} 6(x - 1) + 23$$
$$8x + 16 - 5 \stackrel{?}{=} 6x - 6 + 23$$

Group like terms and simplify.
Move variables to the same side
of the equation.
Subtract 11 from both sides of the equation.
Divide both sides by 2.

$$8x + 11 \stackrel{?}{=} 6x + 17$$
$$\underline{-6x \qquad\quad -6x}$$
$$2x + \cancel{11} \stackrel{?}{=} 17$$
$$\underline{-\cancel{11} \quad -11}$$
$$\frac{\cancel{2}x}{\cancel{2}} \stackrel{?}{=} \frac{6}{2}$$
$$x = 3$$

The ending statement has **one solution**. The only value that makes the equation true is $x = 3$.

DETERMINING THE NUMBER OF SOLUTIONS TO A LINEAR EQUATION

A linear equation in one variable will have one solution, no solution or infinitely many solutions. This can be determined once the equation is in its simplest form.

- One Solution – When solved, the variable is equal to one number.

- No Solution – When solved, the equation makes a false statement (i.e. $5 = 4$).

- Infinitely Many Solutions – When solved, the equation makes a statement that is true no matter what value of the variable is tested (i.e. $12 = 12$, $4x = 4x$).

Lesson 1.7 ~ Solutions to Linear Equations

EXERCISES

1. Write a simple equation that has no solutions.

2. Write a simple equation that has infinitely many solutions.

3. William solved the equation in the box below. How many solutions does his equation have? Explain how you know.

$$\begin{aligned} 2(x+2) &= 2x+4 \\ 2x+4 &= 2x+4 \\ -2x \quad &\quad -2x \\ \hline 4 &= 4 \end{aligned}$$

Solve each equation. Describe the number of solutions (one, none or infinitely many).

4. $2(x + 7) = 2x + 7$

5. $5x - 3 = x + 17$

6. $-3x = 4x - 49$

7. $10 + 6x - 3 = 2x + 7 + 4x$

8. $4(x - 2) + 1 = -7$

9. $\frac{1}{2}(x + 6) = \frac{1}{2}x + 3$

10. $\frac{x}{3} + 12 = 2$

11. $10x - 2 - 6x = 2(2x - 1)$

12. $5x - 30 = 5x + 30$

13. $28 = 12 - 4x$

14. $6x - 1 = 3(2x + 4) - 1$

15. $4x - 0.8 = 3x - 0.8 + x$

16. Tara and Alice joined the same movie club. Tara said her monthly fee is represented by the expression $2x + 8$. Alice said her monthly fee is $2(x + 4)$. In both expressions, x is the number of movies rented in a month. Who got the best deal? Explain how you know.

17. Quinn and Logan solved the equation $8(x - 5) = 8x + 40$. Quinn said the answer was $x = 0$ and Logan said there are no solutions. Who is correct? Support your answer with work.

18. Nancy wrote two equivalent expressions. If she sets them equal to each other, will the equation have one solution, no solution or infinitely many solutions? Use an example to support your answer.

19. Kareem wanted to write an equation for his friend to solve that would have infinitely many solutions. Help him write an equation for his friend. Explain how you know this equation fits what Kareem is looking for.

20. Jared told his teacher that it is impossible to have a one-step equation (only one operation in the equation) that has no solution. Is Jared correct? Explain your reasoning.

REVIEW

Simplify each expression.

21. $9 + 2x - 4 + 8x$

22. $7(x - 2) + 6(x + 1)$

23. $7 + 3(x - 4)$

24. $-9x + 8x + 7y - 6y$

25. $7x + 3y - x + 4y - 2x$

26. $12y - 3x + 10x - y$

Write a simplified expression for the perimeter of each figure.

27.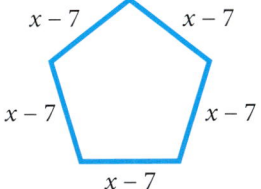
(pentagon with all sides $x - 7$)

28.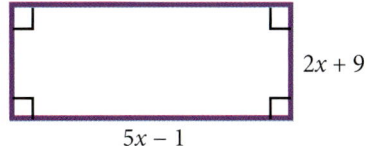
(rectangle with width $2x + 9$ and length $5x - 1$)

Tic-Tac-Toe ~ Small Business Profits

When someone opens a small business, they often use linear equations to determine profits and losses. Three individuals started up three different businesses which are described below.

Business #1
Sarah's Coffee Shop
Start-Up Cost = $275
Profit per Coffee Sold = $2.50
Equation:
$P = -275 + 2.50C$

Business #2
Jason's Skateboarding Store
Start-Up Cost = $400
Profit per Board Sold = $14

Business #3
Jamal's Gaming Shop
Start-Up Cost = $750
Profit per Game Sold = $9.75

1. An equation representing total profits is shown for Business #1. Write the equations for the total profits for the other two businesses.
2. For each business, determine how many items they will need to sell to earn back the amount of money they spent for starting their business.
3. Each business has a goal of making a total profit of $10,000 in the first quarter of the year. Determine how many total items they will each need to sell to reach their goal.
4. Design your own small business.
   a. Choose one item to sell. Describe why you would like to sell this item.
   b. Estimate the total start-up costs. Explain how you came up with this amount.
   c. Estimate the total profit you hope to make per item sold. Explain how you came up with this amount.
   d. Repeat #1 - #3 above for your small business idea.
   e. Do you think there would be enough interest in your product to reach the goal of $10,000 in the first quarter? Explain your reasoning.

LINEAR INEQUALITIES IN ONE VARIABLE

LESSON 1.8

 Solve inequalities with one variable.

Nathan has more than $10 in his wallet. Jackie has run at most 200 miles this year. Each of these statements can be written using an **inequality**. Inequalities are mathematical statements which use >, <, ≥ or ≤ to show a relationship between quantities.

Nathan has more than $10 in his wallet.

$n > \$10$

Jackie has run at most 200 miles.

$j \leq 200$

Inequalities have multiple answers that can make the statement true. In Nathan's example, he might have $20 or $100. All that is known for certain is that he has more than $10 in his wallet. In Jackie's example, she might have run 200 miles this year or 5 miles. There are an infinite number of possibilities that make each statement true.

INEQUALITY SYMBOLS

>   "greater than"
<   "less than"
≥   "greater than or equal to"
≤   "less than or equal to"

EXAMPLE 1

Write an inequality for each statement.
a. Carla's weight (w) is greater than 100 pounds.
b. Vicky has at most $500 in her savings account. Let m represent the amount of money in Vicky's account.
c. Quinton's age is greater than 40 years old. Let a represent Quinton's age.

SOLUTIONS

a. The key words are "greater than". Use the symbol >. $w > 100$

b. The key words are "at most". This means she has less than or equal to $500. Use the ≤ symbol. $m \leq 500$

c. The key words are "greater than". Use the > symbol. $a > 40$

It is not possible to list all of the solutions to an inequality. In **Example 1c**, Quinton could be 41, 42 or even 43.5 years old. All the answers can be shown on a number line.

$a > 40$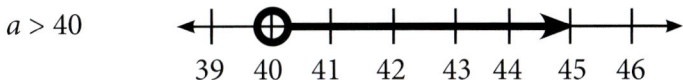

Forty is not included in the solution because Quinton's age is greater than 40, not equal to 40. This is shown on the number line with an "open circle" at 40. All of the numbers to the right of 40 are greater than 40 so they are included in the solution. This is shown with a line and an arrow pointing to the rest of the values.

When using the > or < inequality symbols, an "open circle" is used on the number line because the solution does not include the given number. When using the ≥ or ≤ inequality symbols, a "closed (or filled in) circle" is used because the solution contains the given number. Determining which direction the arrow should point is based on the relationship between the variable and the solution. The arrow points towards the set of numbers that make the statement true.

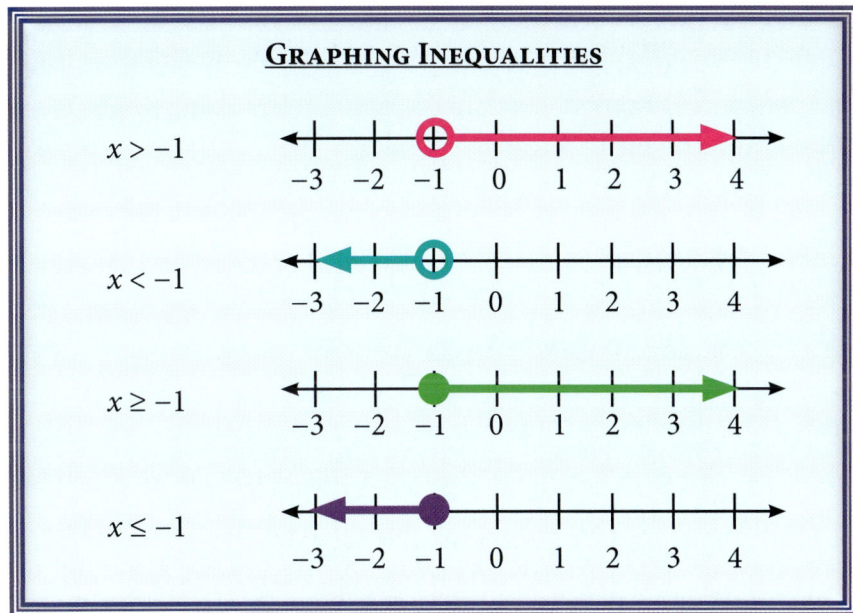

Inequalities are solved using properties similar to those you used to solve equations. Use inverse operations to isolate the variable so the solution can be graphed on a number line.

EXAMPLE 2   Solve the inequality and graph its solution on a number line.
$$\frac{x}{4} + 2 \geq 3$$

SOLUTION   Subtract 2 from both sides of the inequality.

Multiply both sides of the inequality by 4.

$$\frac{x}{4} + 2 \geq 3$$
$$\phantom{\frac{x}{4}} -2 \ -2$$
$$4 \cdot \frac{x}{4} \geq 1 \cdot 4$$
$$x \geq 4$$

Graph the solution on a number line. Use a closed circle.

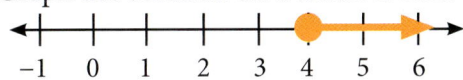

Lesson 1.8 ~ Linear Inequalities in One Variable

EXAMPLE 3 Solve the inequality and graph its solution on a number line.
$$6x + 3 < 2x - 5$$

SOLUTION Subtract $2x$ from each side of the inequality.

Subtract 3 from each side.

Divide both sides by 4.

$$\begin{array}{r} 6x + 3 < 2x - 5 \\ \underline{-2x \quad\quad -2x} \\ 4x + 3 < -5 \\ \underline{-3 \quad -3} \\ \dfrac{4x}{4} < \dfrac{-8}{4} \\ x < -2 \end{array}$$

Graph the solution on a number line. Use an open circle.

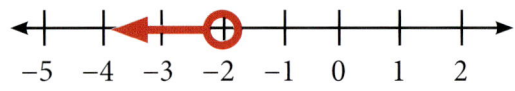

One special rule applies to solving inequalities. Whenever you multiply or divide by a negative number on both sides of the equation, you must flip the inequality symbol. For example, less than (<) would become greater than (>) if you multiply or divide by a negative number.

EXAMPLE 4 Solve the inequality $-4x + 7 \leq 19$.

SOLUTION Subtract 7 from each side of the inequality.

Divide both sides by -4.

Since both sides were divided by a negative, flip the inequality symbol.

$$\begin{array}{r} -4x + 7 \leq 19 \\ \underline{-7 \quad -7} \\ \dfrac{-4x}{-4} \leq \dfrac{12}{-4} \\ x \geq -3 \end{array}$$

The sign changed direction because both sides were divided by a negative number.

EXERCISES

Write an inequality for each graph shown. Use x as the variable.

1.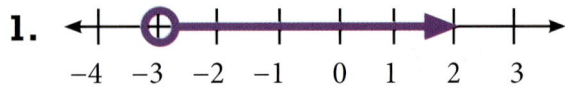

2.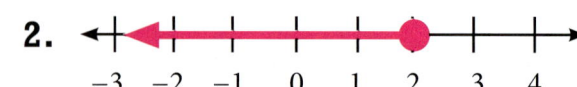

3.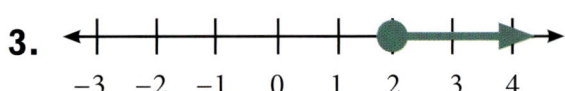

4.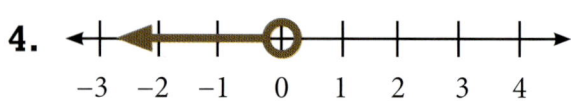

Solve each inequality. Graph the solution on a number line.

5. $4x - 1 \geq 15$ **6.** $10 < 6 + 2x$ **7.** $3x + 10 < -2$

8. $5x - 7 > 8$ **9.** $\frac{x}{2} - 1 \geq -1$ **10.** $-3x - 4 < 5$

11. $3(x + 1) \geq 9$ **12.** $3 < 1 + \frac{x}{-4}$ **13.** $5x < 2x - 21$

14. $-2 + 4x \geq 3 - 6x$ **15.** $2(x + 3) \geq 5x + 12$ **16.** $\frac{1}{2}x + 2 < x$

17. A forklift has a maximum carrying capacity of 960 pounds. Each cargo box weighs 60 pounds.
  a. Write and solve an inequality that represents the maximum number of cargo boxes the forklift can hold.
  b. A 120-pound carrying case is used to hold the cargo boxes. What is the maximum number of cargo boxes the forklift can carry when the carrying case is used? Show that your answer is correct by showing that one more than your answer would exceed the forklift's capacity.

18. Olivia has $700 in her bank account at the beginning of the summer. She wants to have at least $150 in her account at the end of the summer. Each week she withdraws $40 for food and entertainment.
  a. Write an inequality for this situation. Let x represent the number of weeks she withdraws money from her account.
  b. What is the maximum number of weeks that Olivia can withdraw money from her account? Explain how you know your answer is correct.
  c. How much money will be left in her account after the last full withdrawal?

19. Ivan was at the beach. He wanted to spend $12 or less on a beach bike rental. The company he chose to rent from charged an initial fee of $5 and an additional $0.45 per mile he rode.
  a. Write an inequality for this situation. Let x represent the number of miles ridden.
  b. How many miles can Ivan ride without going over his spending limit? Write your answer as a whole number.

20. Mike went to the arcade. He spent less than or equal to $30; but spent more than $25. Create a number line that shows all the possible amounts that Mike may have spent at the arcade. Use words and/or numbers to show how you determined your answer.

REVIEW

Solve each equation. Describe the number of solutions (one, none or infinitely many).

21. $4x - 3 = x + 18$ **22.** $4(x - 1) = 4x - 8$ **23.** $5x - 3x + 2 = 2x + 2$

24. $\frac{1}{2}(x + 4) = \frac{1}{2}x + 2$ **25.** $6(x - 3) = x - 3$ **26.** $9 + 6x = 6x + 2 + 4$

Tic-Tac-Toe ~ Compound Inequalities

Compound inequalities contain two inequality symbols. An example of a compound inequality is $4 < x < 9$. This can be read "x is greater than four and less than 9." A solution to a compound inequality is a value that makes the statement true. In this case, any number between 4 and 9 makes the statement true. For example, 5.5 is a solution, $4 < 5.5 < 9$, because it is greater than 4 but less than 9.

To graph a compound inequality, place open or closed circles (depending on the inequality) on each end value and connect the circles with a line in between.

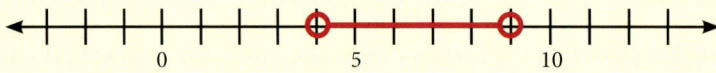

When solving compound inequalities, you must isolate the variable in the middle part of the inequality. In order to maintain the balance of the inequality, you must perform operations on ALL THREE parts of the inequality (left, middle and right)

Example: Solve $2 \leq x - 1 < 11$. Graph the solution.

Add 1 to all three parts of the inequality.

$$\begin{array}{c} 2 \leq x - 1 < 11 \\ +1 \quad\; +1 \;\; +1 \\ \hline 3 \leq x < 12 \end{array}$$

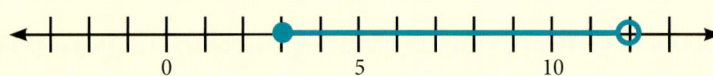

Solve each inequality below. Graph the solution on a number line.

1. $2 \leq x + 2 \leq 6$

2. $6 < 2x < 16$

3. $-5 \leq x - 2 < -3$

4. $5 < 3x + 5 \leq 17$

5. $1 < \frac{x}{3} - 1 < 2$

6. $\frac{1}{2} \leq 2x - 3 \leq 1$

Tic-Tac-Toe ~ One Does Not Belong

Create ten game cards. Each game card needs three equations on it that require two or more steps to solve. Two of the equations on a card should represent equations that have the same number of solutions (one, none or infinitely many). The other equation should have a different number of solutions. Participants try to locate the equation that does not fit. The cards can be used as a game or as a full class activity like a warm-up. Change the placement of the equations that do not belong so they are not always in the same spot on the cards.

38   Lesson 1.8 ~ Linear Inequalities in One Variable

REVIEW

BLOCK 1

 Vocabulary

absolute value	Distributive Property	like terms
algebraic expression	equation	order of operations
coefficient	equivalent expressions	term
constant	inequality	zero pair
	inverse operations	

Find the value of expressions using the order of operations.
Use the order of operations to evaluate expressions.
Simplify expressions using the Distributive Property and combining like terms.
Use inverse operations to solve one-step equations.
Solve two-step equations.
Simplify and solve multi-step equations.
Determine if a linear equation in one variable has no solution, one solution or infinitely many solutions.
Solve inequalities with one variable.

Lesson 1.1 ~ Order of Operations

Evaluate each expression.

1. $6 + 7 \cdot 3 - 9$

2. $35 \div 7 \cdot 2 - 4^2$

3. $|4 - 8| + |8 - 1|$

4. $(-2 + 5)^2 - 10$

5. $2(11 - 5) - 10$

6. $\dfrac{9 + 11}{(1 + 1)^2}$

7. $\dfrac{-14 - 4}{2 + 1} - 5$

8. $\dfrac{7|11 - 4| + 1}{(-5)^2}$

9. $\frac{1}{2}(6 + 8) - 2(3 - 5)$

Lesson 1.2 ~ Evaluating Expressions

Evaluate each expression for the given values of the variables.

10. $2x + 3$ when $x = 4$

11. $12 - 3f$ when $f = 1.2$

12. $\dfrac{2y - 1}{2y}$ when $y = 3$

13. $10(p + 5)$ when $p = -7$

14. $6a^2 - 14$ when $a = 3$

15. $7m + 1$ when $m = \frac{1}{2}$

Block 1 ~ Review 39

Copy each table. Complete each table by evaluating the given expression for the values listed.

16.

x	$5x + 1$	Output
−2		
0		
$\frac{1}{2}$		
3		
7		

17.

x	$\frac{3x + 2}{4}$	Output
−3		
0		
4		
10		

Lesson 1.3 ~ The Distributive Property

Use the Distributive Property to simplify each expression.

18. $5(x + 4)$ **19.** $\frac{1}{4}(8x - 2)$ **20.** $-1(3x - 12)$

21. $6(3x - 10)$ **22.** $-3(-8x + 5)$ **23.** $0.1(15x + 5)$

Simplify each expression.

24. $3x + 7 + 4x - 2$ **25.** $3(x - 4) + 6$ **26.** $5(x - 1) - 5$

27. $3(x - 4) + 2(x + 1)$ **28.** $4x + 3x - 5x + 2 - x$ **29.** $6x + 40 + 4x - 15$

Write and simplify an expression for the perimeter of each figure.

30.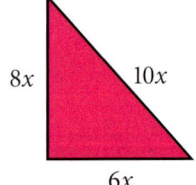

31.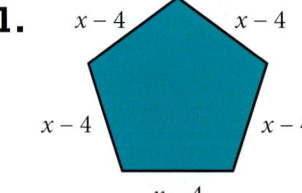

Lesson 1.4 ~ Solving One-Step Equations

Solve each equation. Show all work necessary to justify your answer.

32. $x + 13 = 35$ **33.** $\frac{x}{7} = -3$ **34.** $10x = 90$

35. $-8x = -44$ **36.** $\frac{1}{2}x = 10$ **37.** $3.7 = \frac{x}{6}$

Write an equation for each statement. Solve each equation. Show all work necessary to justify your answer.

38. The sum of seven and a number is 73.

39. Ten less than a number is 66.

40. Twenty-nine is equal to a number divided by three.

Lesson 1.5 ~ Solving Two-Step Equations

Solve each equation. Show all work necessary to justify your answer.

41. $10x - 8 = 92$

42. $-5x - 1 = 44$

43. $4 = \frac{x}{6} - 2$

44. $\frac{x}{7} + 1.3 = 2.1$

45. $2x - 7 = 4$

46. $\frac{1}{3}x + 3 = 3$

47. Kari is saving money for an MP3 player. She began the year with $30 of savings. At the end of each month, Kari adds $14.
   a. How much will Kari have after 3 months have passed?
   b. Write a formula to calculate Kari's total savings (S) based on how many months (m) she has saved this year.
   c. The MP3 player Kari wants costs $170. How many months will it take for Kari to have enough money to purchase the MP3 player? Use mathematics to justify your answer.

Lesson 1.6 ~ Solving Multi-Step Equations

Solve each equation. Show all work necessary to justify your answer.

48. $6(x + 3) = 42$

49. $5x + 2 = 3x - 2$

50. $-4x + 9 = x - 11$

51. $2(x - 1) = 3x - 13$

52. $\frac{1}{4}(8x + 1) = 12\frac{1}{4}$

53. $9x + 4x - 7 = 3x - 17$

54. A bowling alley has two payment options for bowlers. Option A allows bowlers to pay $3 per game. Option B allows bowlers to join the 'Elite Bowling Club' for $15 plus an additional $1.50 per game.
   a. Write an expression for the cost of x games with Option A.
   b. Write an expression for the cost of x games with Option B.
   c. Set the two expressions equal to one another. Solve the equation. How many games would someone need to bowl to make both options cost the same amount?
   d. Ray is going to bowl 14 games; which option should he choose? Explain your reasoning.

55. Julie had $400 in her savings account at the beginning of the summer. Each week she took $20 out of her account. Stephen had $100 in his account at the beginning of summer. Each week he added $30 to his account. After how many weeks did Julie and Stephen have the same amount in their accounts? Use numbers, symbols and/or words to show how you determined your answer.

56. Describe the first step you would take to solve the equation below for x.
$$8(x - 4) + 3 = 4x$$

57. Mark says he can divide both sides of the equation $-5(x + 1.4) = 80$ by -5 first. Jake says Mark needs to distribute before doing any inverse operations. Who is correct? Explain your reasoning.

Lesson 1.7 ~ Solutions to Linear Equations

Solve each equation. Describe the number of solutions (one, none or infinitely many).

58. $2(x + 3) - 5 = 9$

59. $3(x - 5) = 3x - 15$

60. $2x - 5x + 7 = -3x + 4$

61. $\frac{1}{2}(4x - 3) = 2x + 5$

62. $2x + 7 = x + 7$

63. $6x + 10 = 5x + 10 + x$

64. $4(x + 1) = 3 + 4x - 1$

65. $10x - 3 = 4x + 18$

66. $\frac{x}{5} - 5 = 1$

67. The perimeter of one figure is $\frac{x}{2} + 12$. The perimeter of another figure is $\frac{1}{2}(x + 12)$. Joey says their perimeters are equal. Do you agree or disagree? Explain your answer.

Lesson 1.8 ~ Linear Inequalities in One Variable

Solve each inequality. Graph the solution on a number line.

68. $6x + 8 \geq 50$

69. $2 < -1 + 3x$

70. $\frac{x}{-3} - 1 \geq -2$

71. $11x - 7 > 2x + 2$

72. $5 < \frac{x}{5} + 6$

73. $-3(x + 1) \geq 9$

74. A serving bowl used for fruit has a maximum capacity of 18 cups. Each scoop of fruit is $\frac{1}{3}$ cup.
   a. Write and solve an inequality that represents the maximum number of scoops the bowl holds.
   b. Two cups of marshmallows were added to the bowl prior to the fruit being added. What inequality represents the number of scoops the bowl can hold after the marshmallows were added?

Tic-Tac-Toe ~ Equation Mats

Equation mats are used to see a visual model of the equation-solving process. Write a "How To…" guide about using equation mats for different types of equations. Include the following types of equations in your guide:
  • One-Step Equations
  • Two-Step Equations
  • Equations with Variables on Both Sides of the Equals Sign

Tic-Tac-Toe ~ Math in Everyday Life

Create a booklet of eight word problems that can be solved using multiple-step equations and inverse operations. On each page, write a word problem. On the corresponding back side of the page in the booklet, show how to solve the problem using equations. Be sure to end the page by answering the problem in a complete sentence.

CAREER FOCUS

RODGER
INSURANCE AGENCY OWNER

I am the owner of an insurance agency that deals mostly with farmers. We consult with clients to help them understand how to protect themselves from disaster by having the right kind of insurance. We also offer prices for whatever coverage a client may need. If a customer decides to buy insurance, we deliver their policy and explain it to them. Our company tries to give excellent service so our clients will want to stay with us for a long time.

The amount of money a client must pay for insurance is determined by math formulas. Most of the time, calculations are done by computers. Some insurances, though, are still done without computers. One type of insurance where we do the math ourselves is crop insurance. If a farmer tells us he wants to get insurance on 500 tons of hay, we have to determine a couple of things. The first question to answer is how much they want the hay insured for. The next question is for how long they would like the hay covered. There is a table that gives all of the different rates of policies depending on those two questions. We can use the rates from the table to come up with a price for the farmer's policy.

Insurance agents also use math to determine fire insurance rates. An insurance agent has to add all the different costs that would go into rebuilding a house if it were to burn down. Some of these calculations can become pretty complicated. In many ways, it is just like a big story problem.

You do not need a college education to work in the insurance field, but it is a good idea to have one. There are many different career paths in insurance, and a college education will help you in whatever path you choose to follow.

Insurance is commissions-based. This means that you make a percentage of whatever amount of insurance you sell. The more you sell, the more money you make. The harder you work, the more you are rewarded.

I like being an insurance agency owner. As the owner I do a little bit of everything. I fix computers, answer phones, meet with clients and pay the employees and bills. No day is the same, and no day is ever a slow day. I also like the fact that the effort you put into the work affects how much you earn. Not all jobs are like that. Lastly, I like my career because it gives me a very satisfying feeling to know that I helped someone out when they encountered a situation in life that required insurance.

CORE FOCUS ON LINEAR EQUATIONS
BLOCK 2 ~ SEQUENCES AND SLOPE

Lesson 2.1	Recursive Routines	46
	Explore! Caloric Recursive Routines	
Lesson 2.2	Linear Plots	51
Lesson 2.3	Recursive Routine Applications	56
	Explore! Saving and Spending	
Lesson 2.4	Rate of Change	61
Lesson 2.5	Recursive Routines to Equations	67
	Explore! Modeling with Equations	
Lesson 2.6	Input-Output Tables from Equations	73
	Explore! Linear Qualities	
Lesson 2.7	Calculating Slope from Graphs	77
Lesson 2.8	The Slope Formula	82
	Explore! Find That Formula	
Review	Block 2 ~ Sequences and Slope	86

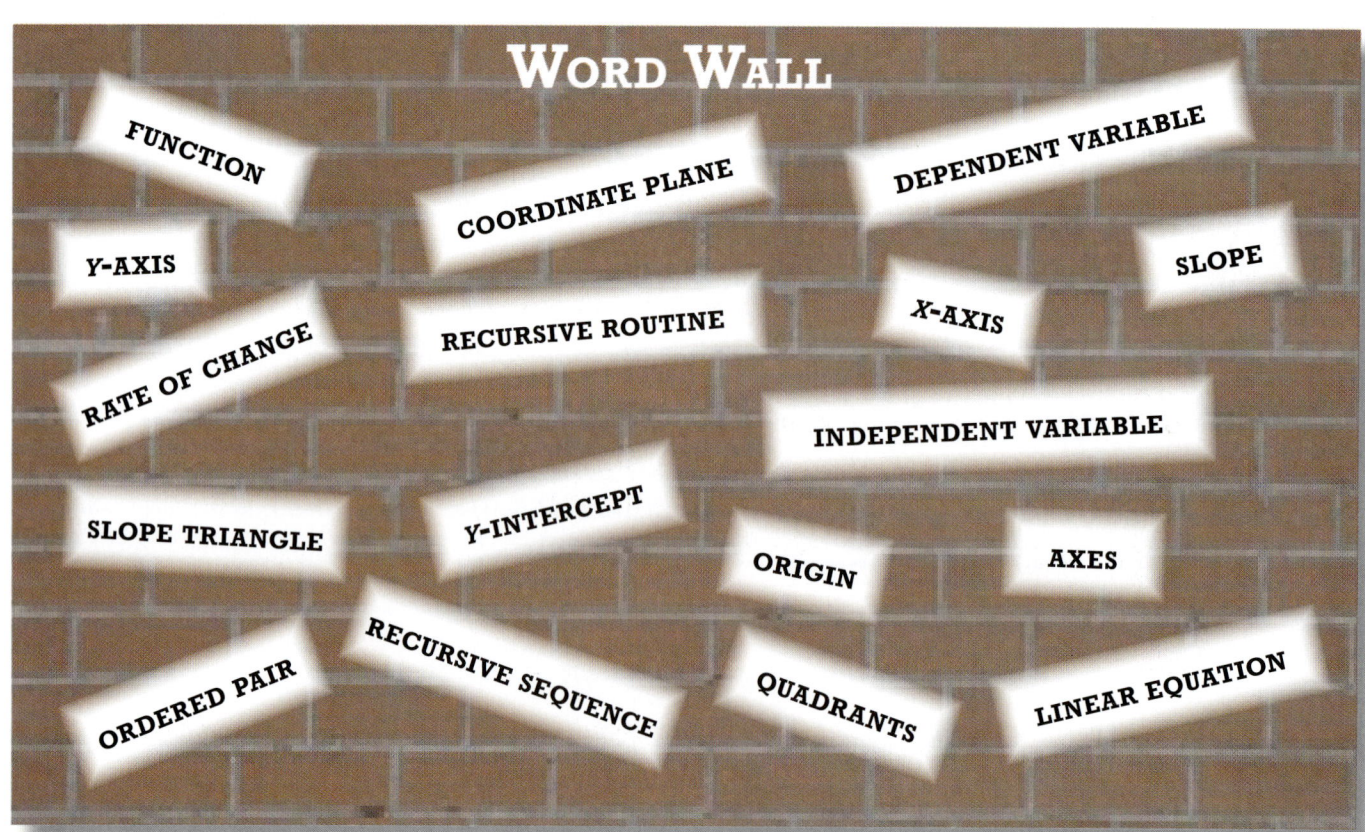

BLOCK 2 ~ SEQUENCES AND SLOPE TIC-TAC-TOE

SLOPE METHODS

Create a flip book explaining how to find slope from tables, graphs and ordered pairs.

See page 85 for details.

CROSSING PATHS

Determine where two recursive sequences cross paths. Illustrate solutions with tables and graphs.

See page 50 for details.

CARD GAME

Make a card game where players create recursive routines and score points.

See page 60 for details.

GEOMETRIC SEQUENCES

Examine recursive sequences involving repeated multiplication. Write equations for these recursive routines.

See page 72 for details.

CHALLENGING TABLES

Find the rates of change and start values in challenging input-output tables.

See page 66 for details.

WRITING EQUATIONS FROM TABLES

Create a worksheet to help another student through the process of writing an equation for an input-output table.

See page 76 for details.

RATE APPLICATIONS

Find the rates of change in real-world situations. Write application problems.

See page 66 for details.

CHILDREN'S STORY

Write a children's story about recursive sequences. The main character encounters a recursive sequence and develops its equation.

See page 81 for details.

SIMILAR SLOPE TRIANGLES

Make discoveries about different-sized slope triangles formed on the same line.

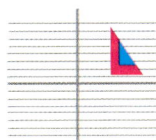

See page 91 for details.

RECURSIVE ROUTINES

LESSON 2.1

 Write recursive routines and create recursive sequences.

A **recursive sequence** is an ordered list of numbers that begins with a start value. Each term in the sequence is generated by applying an operation to the term before it. This same operation is repeated to the resulting value. This process continues to make a sequence of terms.

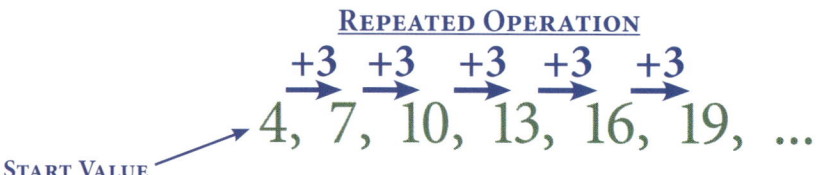

A **recursive routine** is described by stating the start value and the operation that is performed to get to the next term. In this case, the recursive routine for the sequence above is:

> Start Value: 4
>
> Operation: Add 3

EXAMPLE 1 For each of the following recursive sequences, state the start value, operation and the next three terms.

a. 8, 17, 26, 35, 44, …

b. 6, 2, −2, −6, −10, …

SOLUTIONS

a. Start Value = 8
Operation = Add 9
Next three terms: 53, 62, 71

b. Start Value = 6
Operation = Subtract 4
Next three terms: −14, −18, −22

EXPLORE! — CALORIC RECURSIVE ROUTINES

Step 1: Resting Metabolic Rate (RMR) represents the number of calories your body burns daily when at rest. In the table below, choose the weight and gender that best describes you to determine your approximate RMR. Record your value on your paper.

MALE

Weight (lbs)	RMR (kcal)
80	1290
90	1340
100	1400
120	1490
140	1600
160	1720
180	1830

FEMALE

Weight (lbs)	RMR (kcal)
80	1130
90	1170
100	1230
120	1320
140	1430
160	1550
180	1660

EXPLORE! (CONTINUED)

Step 2: Choose an activity that you would most like to participate in from the list below. Record your choice and the calories burned per minute.

Activity	Aerobics	Downhill Skiing	Bowling	Horseback Riding	Flag Football
Calories Burned per Minute	7	6	3	4	8

Step 3: Copy the table shown at right. Insert the name of your activity at the top of the first column.

Step 4: How many calories has your body burned during a full day before you participate in your chosen activity? What is this value called? Where would this fit in the table?

Step 5: Determine the total daily calories burned through the first five minutes of your activity. Continue your calculations to determine the total daily calories burned for 10 minutes, 20 minutes and 30 minutes.

Minutes Spent (Insert activity)	Total Daily Calories Burned
0	
1	
2	
3	
4	
5	
10	
20	
30	
	≈ 2,000

Step 6: Say you want to burn 2,000 total calories during one day. How many minutes will you need to participate in your activity? Is this reasonable?

Step 7: Describe the recursive routine for your table (when going up one minute at a time) by giving the start value and the operation that must be performed to arrive at the next term.

RECURSIVE ROUTINE
Start Value: _____
Operation: _____

EXAMPLE 2

Find the missing values in each sequence. Identify the start value and the operation that must be performed to arrive at the next term.
a. 25, 19, 13, _____, 1, _____, _____   b. 32, 45, _____, _____, 84, _____

SOLUTIONS

a. The numbers in the list are going DOWN 6 each time.
Start Value: 25
Operation: Subtract 6
Completed List: 25, 19, 13, <u>7</u>, 1, <u>−5</u>, <u>−11</u>

b. The numbers in the list are increasing by 13 each time.
Start Value: 32
Operation: Add 13
Completed List: 32, 45, <u>58</u>, <u>71</u>, 84, <u>97</u>

Lesson 2.1 ~ Recursive Routines   47

USING YOUR CALCULATOR TO CREATE A RECURSIVE ROUTINE

Most scientific and graphing calculators can perform a repeated calculation to create a sequence of numbers.
- Enter the start value.
- Press ENTER or =.
- Enter the operation.
- Press ENTER or = repeatedly to generate the recursive sequence.

EXAMPLE 3

For each sequence, describe the recursive routine by giving the starting value and operation. List the next three terms.

a. 35, 50, 65, 80, …
b. 10, 1, −8, −17, …

SOLUTIONS

a.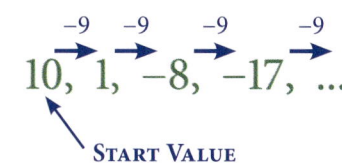

Start Value: 35
Operation: Add 15 (or +15)
Next three terms: 95, 110, 125

b.

Start Value: 10
Operation: Subtract 9 (or − 9)
Next three terms: −26, −35, −44

EXERCISES

Copy each sequence of numbers and fill in the missing values. Identify the start value and the operation that must be performed to arrive at the next term.

1. 3, 5, ____, 9, ____, ____

2. 125, ____, 175, 200, ____, ____

3. 27, 22, 17, ____, ____, ____

4. ____, ____, −5, −9, −13, ____

5. 23, ____, 45, 56, ____, ____

6. $4\frac{1}{2}$, 5, ____, ____, $6\frac{1}{2}$, ____

7. ____, 2.7, 3.1, ____, 3.9, ____

8. −42, −22, ____, ____, ____, 58

9. Deidre weighs 100 pounds. She enjoys downhill skiing at Alpine Meadows.

Minutes Spent Downhill Skiing	Total Daily Calories Burned
0	
1	
2	
3	
4	

   a. Copy the table to the right. Use the table in the **Explore!** to determine Deidre's resting metabolic rate (RMR). Insert it in the table for 0 minutes skiing.
   b. How many calories will Deidre burn each minute she skis?
   c. Fill in the total calories she will burn in a full day for each added minute she skis.
   d. Deidre ends up skiing for 40 minutes. How many TOTAL calories did she burn on that day if she participated in no other activities? Use mathematics to justify your answer.

10. Squares, each 1 centimeter by 1 centimeter, are placed next to each other (one at a time) to form a long strip of squares.

    a. What is the perimeter of the first figure using just one square?
    b. What is the perimeter of the figure using two squares? Three squares?
    c. Draw the next two figures in the pattern. What are the perimeters of these figures?
    d. Write the recursive routine (start value and operation) that describes the perimeters.
    e. What is the perimeter of the figure with 14 squares in one strip? Use words and/or numbers to show how you determined your answer.

For each sequence describe the recursive routine (start value and operation). List the next three terms in the sequence.

11. 8, 16, 24, 32 …

12. 10.5, 9.4, 8.3, 7.2 …

13. 9, 5, 1, −3 …

14. $\frac{1}{3}$, 1, $1\frac{2}{3}$, $2\frac{1}{3}$ …

15. Each block at right is 1 centimeter by 1 centimeter.
    a. Draw the next two figures in the following pattern.
    b. What is the perimeter of the first figure using just one square?
    c. What is the perimeter of the second figure? The third figure?
    d. Write the recursive routine (start value and operation) that describes the perimeters.
    e. What is the perimeter of the seventh figure in this pattern? Use words and/or numbers to show how you determined your answer.

16. Mary grew strawberries this summer and plans to open a fruit stand to sell them. On the first day, she will charge $4.50 for a pound of strawberries. Each day that the fruit stand is open, the price of the berries will go down $0.30.
    a. Write a recursive routine that describes the cost of the strawberries.
    b. Write the sequence of numbers that shows the price of strawberries each day for the first five days the fruit stand is open.
    c. On what day will the price of the strawberries drop below $2.00?
    d. When will she be giving away the strawberries for free?

17. Generate your own recursive sequence. Describe the recursive routine by giving the start value and operation. What is the 20th term in your sequence? Use mathematics to justify your answer.

REVIEW

Copy each table. Complete each table by evaluating the given expression for the values listed.

18.
x	6x + 7	Output
−2		
0		
1/2		
5		
10		

19.
x	3(x − 1)	Output
−3		
0		
2.2		
8		
21		

Tic-Tac-Toe ~ Crossing Paths

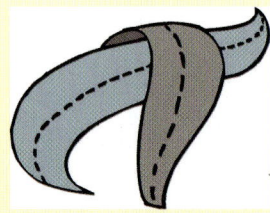

Recursive routines that represent linear relationships often cross paths or intersect at one point. To find where the sequences intersect, you can create an input-output table or make a graph. Each set of recursive routines below will intersect at one point. Find the point of intersection.

Use an input-output table to determine where the two recursive routines intersect.

1. **Routine A**
   Start Value = 7
   Operation = +2

   Routine B
   Start Value = 19
   Operation = −4

2. **Routine A**
   Start Value = −9
   Operation = +7

   Routine B
   Start Value = 19
   Operation = +3

3. **Routine A**
   Start Value = 120
   Operation = −12

   Routine B
   Start Value = −80
   Operation = +28

See **Lesson 2.6** for information on input-output tables and graphing.

Use a graph to determine where the two recursive routines intersect.

4. **Routine A**
   Start Value = 4
   Operation = +3

   Routine B
   Start Value = 20
   Operation = −1

5. **Routine A**
   Start Value = −2
   Operation = +5

   Routine B
   Start Value = 13
   Operation = +2

6. Which method did you prefer for finding the point of intersection? Why?

LINEAR PLOTS

LESSON 2.2

 Create linear plots for recursive sequences.

The **coordinate plane** is created by drawing two number lines which intersect at a 90° angle. These two lines are called the **axes**. Each number line intersects the other at zero. The point where the two lines cross is called the **origin**. The horizontal axis is used for the variable x (called the **x-axis**) and the vertical axis is used for the variable y (called the **y-axis**). The axes divide the coordinate plane into four quadrants. The **quadrants** are numbered I, II, III and IV starting in the top right quadrant and moving counter-clockwise.

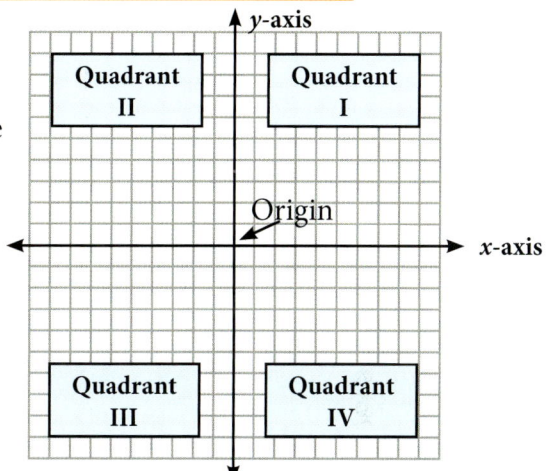

Each point on the graph is named by an **ordered pair**. The first number in the ordered pair corresponds to the numbers on the x-axis. The second number corresponds to the number on the y-axis.

Aroldo attended the State Fair with his friends in August. The entry fee was $8 and each ride he went on cost an additional $2. The graph below shows the total Aroldo may have spent depending on the number of rides he went on.

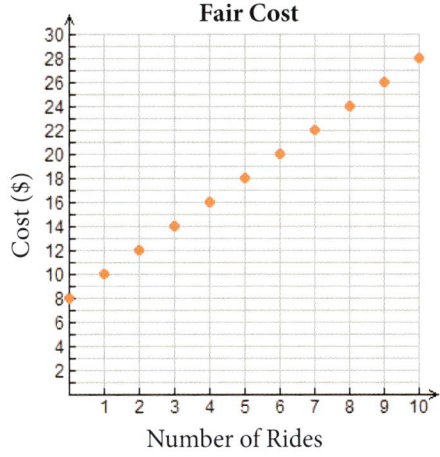

This situation can also be shown using a table. Take each ordered pair on the graph and put it in the corresponding spot in the table.

Number of Rides	0	1	2	3	4	5	6	7	8	9	10
Cost ($)	8	10	12	14	16	18	20	22	24	26	28

The cost (y-coordinate) based on the number of rides can be described by a recursive routine:

> Start Value: $8
> Operation: Add $2

Notice that the points on the graph form a straight line. The graph shows that there is a linear relationship between the number of rides he went on and his total cost. Aroldo's total cost at the fair is directly related to the number of rides he went on. When real-life situations are examined mathematically you will find that many can be described as having a linear relationship. Can you think of any other situations that might have a starting value and then go up or down in equal steps?

In a linear relationship, the y-coordinates follow a recursive routine when the x-coordinates in a table or scatter plot increase in equal increments. When a graph or table shows the x-coordinate increasing by 1, the recursive sequence of the y-coordinates describes the linear relationship.

EXAMPLE 1 Describe the linear relationship given by the y-coordinates on each graph by stating the recursive routine and the first 10 numbers in the recursive sequence.

a.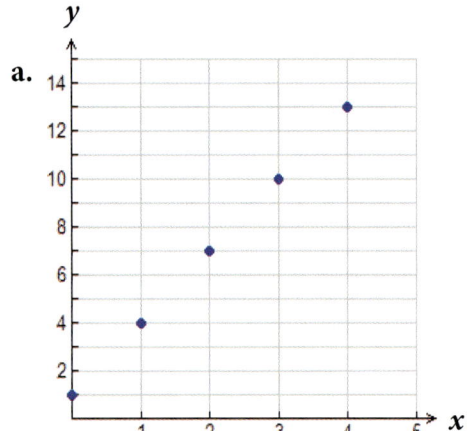

b.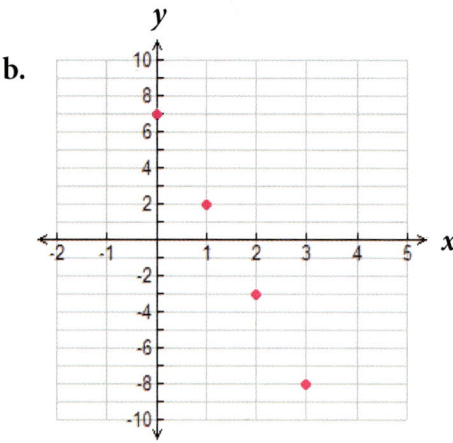

SOLUTIONS

a. The y-coordinates that are shown are 1, 4, 7, 10, 13 …
The recursive routine can be described by the following rules:
Start Value: 1
Operation: Add 3
Using this recursive routine you can determine that the first 10 y-coordinates in this sequence are: 1, 4, 7, 10, 13, 16, 19, 22, 25, 28.

b. The y-coordinates that are shown are 7, 2, −3, −8 …
The recursive routine can be described by the following rules:
Start Value: 7
Operation: Subtract 5
Using this recursive routine you can determine that the first 10 y-coordinates in the sequence are: 7, 2, −3, −8, −13, −18, −23, −28, −33, −38.

Sometimes it is useful to generate a table to represent the ordered pairs of recursive sequences shown on a linear plot. This can be done by creating an input-output table for the *x*- and *y*-coordinates. For example, the ordered pairs of **part a** in **Example 1** can be converted to a table by recording each *x*-coordinate with its corresponding *y*-coordinate.

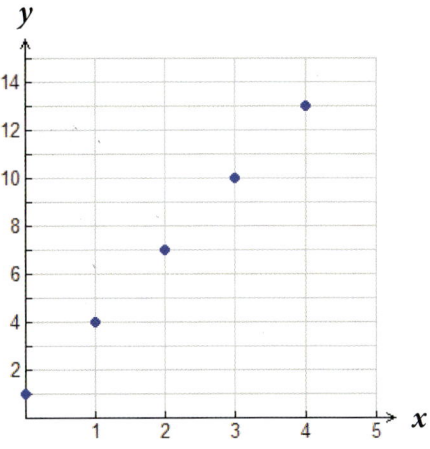

Input *x*	Output *y*
0	1
1	4
2	7
3	10
4	13

Input-output tables can be written horizontally or vertically.

Input *x*	0	1	2	3	4
Output *y*	1	4	7	10	13

EXAMPLE 2

Jerome throws the shot put for Newbridge High School. Coming into the season, his personal best was 42 feet. Each week, his shot put throws increase by 0.5 feet. Create a linear plot that represents this situation. Write a recursive routine to describe it.

SOLUTION

Start Value: 42 Feet
Operation: Add 0.5 Feet

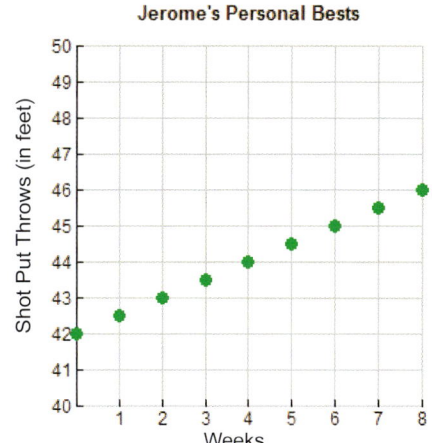

RECURSIVE ROUTINES FOR LINEAR RELATIONSHIPS

Start Value: The *y*-value that corresponds to an *x*-value of 0.
Operation: The amount the *y*-value increases or decreases for each unit on the *x*-axis.

Lesson 2.2 ~ Linear Plots **53**

EXERCISES

Describe the linear relationship given by the y-coordinates on each linear plot by stating the start value and operation. Create an input-output table showing the ordered pairs on each linear plot.

1.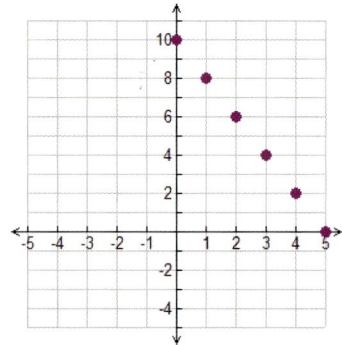

2.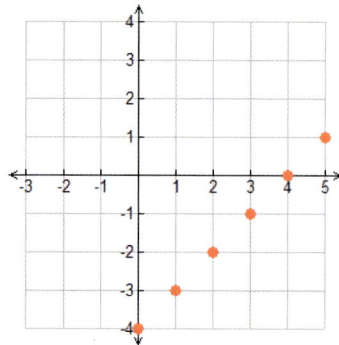

3.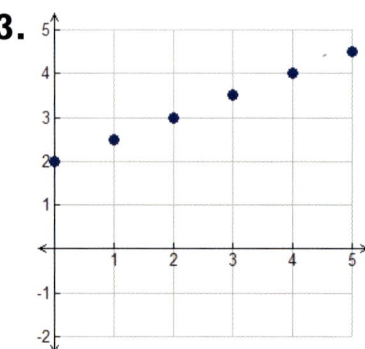

4.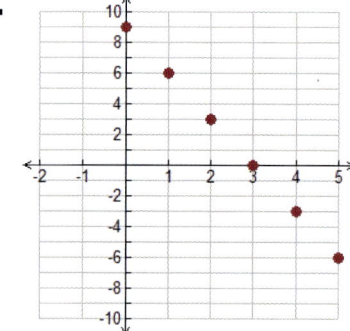

5.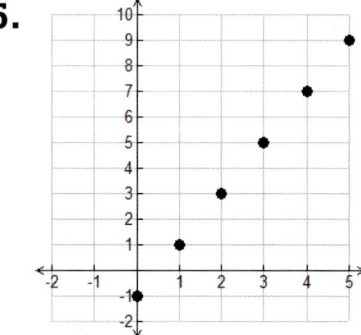

6.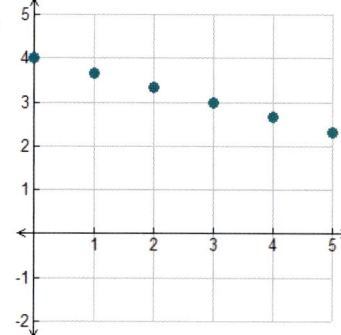

7.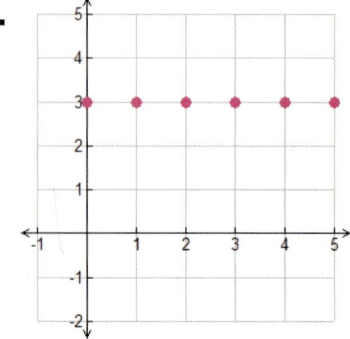

8.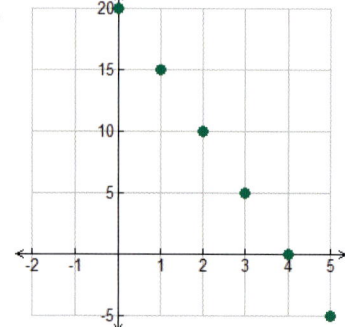

54   *Lesson 2.2 ~ Linear Plots*

9. Kirk chose to go down the steepest, fastest water slide in the aquatic park. The linear plot shows how high off the ground Kirk is, based on the number of seconds he has been on the slide.

   a. Copy the input-output table below and fill in all the ordered pairs shown on the linear plot.

Time (seconds), x	Feet off the Ground, y
0	
1	
2	

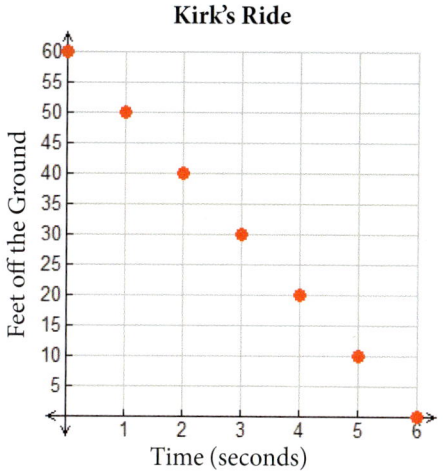

Kirk's Ride

   b. Write the recursive routine for Kirk's ride down the slide.
   c. What does the start value represent in real life?
   d. How long does it take for Kirk to get to the bottom of the slide? Explain how you found your answer.

Create a linear plot for the first five ordered pairs of the given recursive routine.

10. Start Value: 8
    Operation: Subtract 3

11. Start Value: 1
    Operation: Add $\frac{1}{2}$

12. Create a recursive routine by picking a start value and operation (adding or subtracting).
    a. Record your recursive routine on your paper.
    b. Create a linear plot for at least five points in your recursive sequence.
    c. Create an input-output table that shows the x- and y-coordinates for the linear plot you generated.

REVIEW

Write a simplified expression for the area of each figure.

13.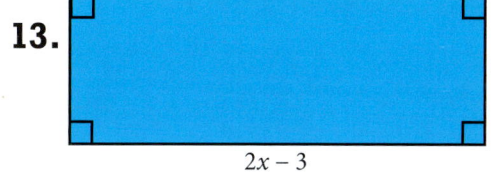

14.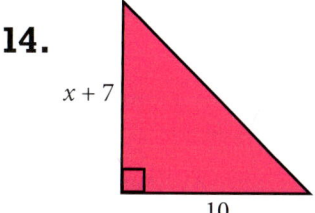

15.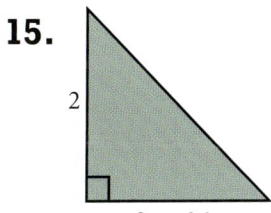

16.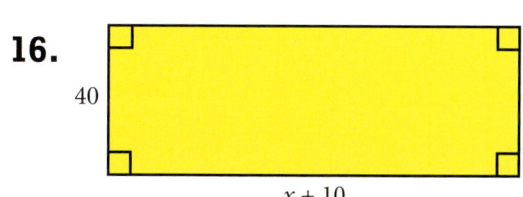

Lesson 2.2 ~ Linear Plots   55

RECURSIVE ROUTINE APPLICATIONS

LESSON 2.3

 Represent recursive routine applications with graphs, tables and words.

Recursive routines are useful when dealing with a variety of real-world situations. Recursive routines can be illustrated with graphs, tables and by words. Using multiple ways of showing a recursive routine helps to reach a variety of audiences. It is important to think about what type of graphic (table, graph, words, etc.) best illustrates each situation.

EXPLORE! SAVING AND SPENDING

William and his sister, Jennifer, each worked summer jobs. William mowed lawns in his neighborhood. Jennifer baby-sat for two different families. By the end of the summer, William had put $410 in a savings account. Jennifer put $275 in her account. After school started, Jennifer continued baby-sitting and earned $20 per week. She put all of her earnings in her savings account. William stopped working and withdrew $15 per week from his savings account for spending money.

Step 1: Write a recursive routine (start value and operation) for the amount in William's savings account each week after school begins.

Step 2: Write a recursive routine for the amount in Jennifer's savings account each week after school begins.

Step 3: Copy the input-output tables shown below and fill in each for the first 10 weeks after school starts.

Weeks After School Starts	William's Total Savings
0	
1	
2	

Weeks After School Starts	Jennifer's Total Savings
0	
1	
2	

Step 4: On the SAME first-quadrant coordinate plane, graph William and Jennifer's total savings for the first ten weeks. Use ● to designate Jennifer's amounts and ▲ to represent William's amounts. Put weeks on the *x*-axis and total savings in dollars on the *y*-axis.

Step 5: After what week does Jennifer have more money than her brother? Which illustration (table, graph or recursive routine) best shows this?

EXAMPLE 1

Matt pays a fee of $25 per month for his cell phone plan. He is charged $0.15 per text message he sends or receives.
a. Write a recursive routine that describes Matt's monthly cell phone bill based on the number of text messages he sent or received.
b. Create an input-output table for the first ten text messages.
c. Create a linear plot that shows his total monthly bill for up to ten text messages.
d. Determine Matt's total bill for the month of January if he sent or received 16 text messages.

SOLUTIONS

a. Start Value = $25
   Operation = Add $0.15 (or + 0.15)

b.

Text Messages Sent or Received	Total Bill
0	$25
1	$25.15
2	$25.30
3	$25.45
4	$25.60
5	$25.75
6	$25.90
7	$26.05
8	$26.20
9	$26.35
10	$26.50

Add $0.15 for each text message received.

c.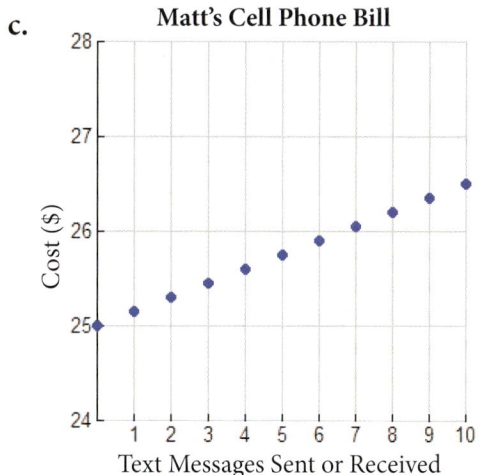

Matt's Cell Phone Bill

d. When using a calculator, enter the start value, 25, and press ENTER or =. Then enter your operation, + 0.15, and press ENTER or = sixteen times. You should arrive at the answer of $27.40.

Lesson 2.3 ~ Recursive Routine Applications **57**

A few things to remember when creating tables and graphing:

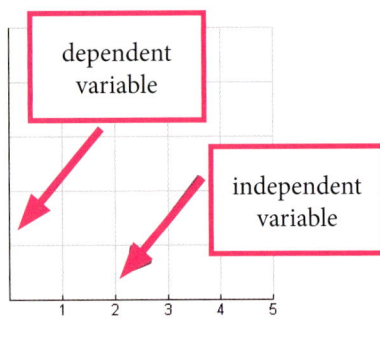

- Always put the **independent variable** in the first column of a table and on the *x*-axis of a graph. The independent variable is the variable the *y*-value is dependent on. The *y*-axis corresponds to the **dependent variable**.

- Most real-world situations take place in the first quadrant. Think about your situation before graphing and decide if negative numbers would ever make sense. For example, you will not have a negative amount for the cost of a cell phone bill, so you will only need to use the first quadrant.

- Choose a range (lowest to the highest number) for the *y*-axis that allows the viewer of your graph to see all points easily. Also, make sure your increments on the *y*-axis are reasonable.

EXERCISES

Determine an appropriate range for the *y*-axis. State what increments you would use on the graph.

1.

Minutes	Distance Traveled
0	8
1	20
2	32
3	44
4	56
5	68

2.

Sales Made	Salary
0	$120
1	$150
2	$180
3	$210
4	$240
5	$270

3.

Years	Car's Worth
0	$12,000
1	$10,500
2	$9,000
3	$7,500
4	$6,000
5	$4,500

4. Jackson got his driving license one year ago. When Jackson got his driver's license, his car insurance cost $82 per month. Each time he gets a speeding ticket, his insurance goes up $26 per month.
   a. Write a recursive routine that describes Jackson's monthly car insurance bill based on the number of tickets he has received.
   b. Create an input-output table for 0 to 5 speeding tickets.
   c. Create a linear plot that shows his total monthly bill through the first five tickets.
   d. Jackson has received 12 speeding tickets. How much is his monthly bill? Show all work necessary to justify your answer.

5. Maggie bought a laptop computer for $799. Each year, the value of her laptop decreases by $70.
   a. Write a recursive routine that describes the value of Maggie's laptop based on the number of years she has owned it.
   b. Create an input-output table for the value of the laptop for 0 to 5 years.
   c. Create a linear plot that shows the value of the laptop through the first five years.
   d. How many years will it take before the laptop is not worth anything? Use words and/or numbers to show how you determined your answer.

6. Frank borrowed $200 from his parents to buy a mountain bike. Each week, he uses $14 of his allowance to pay back his parents.

   a. Write a recursive routine that describes the total amount Frank owes his parents based on the number of weeks that have passed since he borrowed the money.
   b. Create an input-output table that shows the amount he still owes for 0 to 5 weeks.
   c. Create a linear plot that shows the amount Frank still owes his parents through the first five weeks.
   d. How much will Frank's last payment be? Show all work necessary to justify your answer.

7. Quincy hiked up a slope in Desert Shores, California (one of the few places below sea level in the United States). He began at an elevation 61 feet below sea level. Each minute that he hiked, he rose 7 feet in elevation.
   a. Write a recursive routine that describes Quincy's elevation based on the number of minutes he hiked.
   b. Create an input-output table to find his elevation for 0 to 10 minutes of hiking.
   c. Create a linear plot that shows Quincy's change in elevation through the first 10 minutes.
   d. How many minutes did it take for Quincy to get above sea level? Support your answer with mathematics.

8. Victor and Mike had a pizza-eating contest. Victor had already eaten three pieces when the competition started. Mike had only eaten one piece. Once the competition started, Victor was able to eat $\frac{1}{2}$ of a piece every minute. Mike was able to eat a little faster. He ate $\frac{3}{4}$ of a piece every minute.

   a. Write two recursive routines, one that describes Victor's pizza-eating and the other describing Mike's pizza-eating. Label them accordingly.
   b. The pizza-eating competition lasted for 8 minutes. Create two input-output tables that show the number of pieces each boy had eaten for each of the first 8 minutes.
   c. Who won the competition at the end of 8 minutes?

9. When Kathy was born, her grandparents started an account for her college education with $1,000 in it. Each year, on her birthday, they add $250.
   a. Write a recursive routine that gives the amount of money in Kathy's account based on her age, not including interest.
   b. Determine the total amount her grandparents will have contributed after her 18th birthday.
   c. Overall, the entire account earned 28% interest. Determine the total amount the account was worth when she withdrew it after her 18th birthday. Show all work necessary to justify your answer.

10. Marin and Jimmy's father asked them to figure out a problem about an antique desk he owned. The antique desk was currently worth $200. Each year that passed, it was worth $45 more. He wanted to know how many years it would take before the desk would be worth over $500. They created a recursive sequence shown below:

$$200, 245, 290, 335, 380, 425, 470, 515$$

Jimmy says it will take 8 years but Marin says it will only take 7 years. Who do you agree with? Explain your reasoning.

11. Diana's resting metabolic rate is 1,550 calories per day. She participates in a 30-minute aerobics class in the morning and burns calories at a rate of 7 calories per minute. After school she rides her horse. If she burns 4 calories per minute while riding, how long will she need to ride her horse to burn a total of 2,000 calories? Show all work necessary to justify your answer.

REVIEW

Find the missing values in each sequence. Identify the start value and the operation that must be performed to arrive at the next term.

12. −14, −11, _____, _____, −2, _____

13. 5.8, 4.6, _____, 2.2, _____, _____

14. 9, _____, 21, _____, 33, _____

15. $\frac{1}{3}$, 1, _____, $2\frac{1}{3}$, _____, _____

Solve each equation. Show all work necessary to justify your answer.

16. $x + 28 = 102$

17. $\frac{x}{6} = -7$

18. $-3x + 5 = 38$

19. $5x + 7 = 7x - 9$

20. $3 = \frac{x}{2} - 1.5$

21. $2x + 7 = 4$

Tic-Tac-Toe ~ Card Game

Use a regular deck of playing cards for this activity. Take out all the face cards (Jacks, Queens and Kings). Create a card game that can be played with two people. The card game must make the players use recursive routines to earn "points". Be creative with your rules.

Some ideas to think about:

- How many cards does each person start with?
- Do some colors and/or suits represent negative integers or subtraction?
- How will the cards be used to determine a start value of a recursive routine?
- How will the cards be used to determine an operation of a recursive routine?
- How are points scored or how does the person progress towards a finish line?
- Do players ever have to find a specific term in the sequence based on a number they draw from the deck?

Once you have designed your game, ask two different pairs of people to try it out. Ask each player to write a short review of your game once they have played it. Read the reviews and write a one-page paper summarizing the feedback. Also include in your paper any changes you would make in the rules before the game was played again. Turn in your paper along with the original set of rules for your game.

RATE OF CHANGE

LESSON 2.4

 Calculate rates of change and start values.

So far in **Block 2**, you have been looking at many different recursive routines. Each recursive sequence you have examined represents a linear relationship. This means that when you plot the points of the sequence on a coordinate plane, the points fall into a straight line. For each recursive routine, you have been able to define the operation that allows you to move from one number in the sequence to the next because you have been told how much to increase or decrease for each step.

There are many situations where the operation is not given for just one step. For example:

> Colton eats 560 calories in 10 minutes.
> Luke paints 72 pictures in 3 days.
> Karen goes down 90 steps in 4 minutes.

In order to determine the operation needed for each situation, you must calculate the **rate of change**. The rate of change can be found by calculating the change in the output (or y-values) divided by the change in input (x-values). It is also called the unit rate. It is important to think about which term is being used as the 'counter' and place that term in the denominator of the rate because it is your x-value. Time is the most common 'counter'.

You must also determine if a situation is giving you increasing numbers in the recursive sequence (pictures painted per day, calories eaten per minute) or decreasing numbers in the sequence (steps descended per minute). This will help in deciding if you are adding or subtracting your "rate of change" amount.

Situation	Rate	Rate of Change	Operation
Colton eats 560 calories in 10 minutes.	$\dfrac{560 \text{ calories}}{10 \text{ minutes}}$	$\dfrac{56 \text{ calories}}{1 \text{ minute}}$	Add 56
Luke paints 72 pictures in 3 days.	$\dfrac{72 \text{ pictures}}{3 \text{ days}}$	$\dfrac{24 \text{ pictures}}{1 \text{ day}}$	Add 24
Karen goes down 90 steps in 4 minutes.	$\dfrac{90 \text{ steps}}{4 \text{ minutes}}$	$\dfrac{-22.5 \text{ steps}}{1 \text{ minute}}$	Subtract 22.5

INCREASING? *DECREASING?*

EXAMPLE 1 Determine the rate of change for each situation. State the operation that would occur in the recursive routine.
a. Jessica loses $580 in 20 days in the stock market.
b. Patrick earned $53 for delivering 10 packages.

SOLUTIONS a. The independent variable is the number of days. This term goes in the denominator.

$$\frac{-\$580}{20 \text{ days}} \xrightarrow{\div 20} \frac{-\$29}{1 \text{ day}}$$

She is losing money, so the operation involves subtraction.
Operation = Subtract $29

b. The independent variable is the number of packages.

$$\frac{\$53}{10 \text{ packages}} \xrightarrow{\div 10} \frac{\$5.30}{1 \text{ package}}$$

He is earning money, so the operation involves addition.
Operation = Add $5.30

In some situations, information will be given to you in an input-output table. In those cases, you must be able to locate numbers on the table that will allow you to determine the rate of change.

RATE OF CHANGE

The rate of change is the change in y-values over the change in x-values: $\dfrac{\text{change in } y\text{-values}}{\text{change in } x\text{-values}}$

Once the rate of change is determined, locate or calculate the start value from the table. The start value is the y-value that is paired with the x-coordinate of zero.

EXAMPLE 2 Determine the rate of change and start value for the input-output table.

x	y
−1	1
0	4
1	7
2	10
3	13

SOLUTION Choose two ordered pairs. Look for two consecutive numbers in the independent variable column.

Change in x-values = +1   Change in y-values = +3

x	y
−1	1
0	4
1	7
2	10
3	13

Calculate the rate of change. $\dfrac{\text{change in } y\text{-values}}{\text{change in } x\text{-values}} = \dfrac{+3}{+1} = +3$

The start value is 4 because it is the y-value that is paired with an x-value of 0.

In some tables the *x*-coordinate of 0 is not listed. This means that the start value is not given. In order to find the start value you must first find the rate of change. Use the rate of change to work forward or backward to find the *y*-value that is paired with 0.

EXAMPLE 3

The rate of change in the table is −2. Find the start value.

x	y
2	6
3	4
4	2
5	0

SOLUTION

Rewrite the table to include *x*-coordinates to 0.

x	y
0	
1	
2	6
3	4

The rate of change is −2. Work backwards to get to the *x*-coordinate of 0 by doing the opposite of the rate of change. Add 2 for each step.

x	y
0	10
1	8
2	6
3	4

+2
+2

The start value is 10.

EXAMPLE 4

Find the start value and rate of change for the input-output table.

x	y
−2	−9
3	11
7	27
9	35
12	47

SOLUTION

Find the rate of change by selecting two pairs of numbers. Find the change in *x* and the change in *y*.

Any combination of two ordered pairs can be used to find the rate of change.

x	y
−2	−9
3	11
7	27
9	35
12	47

Change in *x*-values = +2   Change in *y*-values = +8

Calculate the rate of change.

$$\frac{\text{change in } y\text{-values}}{\text{change in } x\text{-values}} = \frac{+8}{+2} = +4$$

Lesson 2.4 ~ Rate of Change   **63**

EXAMPLE 4
SOLUTION
(CONTINUED)

Use the rate of change to work forwards from $x = -2$ to find the y-value paired with the x-coordinate of 0.

x	y
−2	−9
−1	−5
0	−1

+4
+4

The start value is −1.
The rate of change is +4.

EXERCISES

Determine the rate of change for each situation.

1. George collected 18 bugs in 9 days.

2. Over 6 days Theo spent $336.

3. Michiko took 760 steps during a 15 minute run.

4. Natalie spent $4.80 for 8 roses.

Determine the rate of change and start value for each table.

5.
x	y
0	5
1	9
2	13
3	17
4	21

6.
x	y
−2	12
−1	9
0	6
1	3
2	0

7.
x	y
0	−3
3	0
5	2
6	3
8	5

8.
x	y
−2	−4
1	2
3	6
5	10
8	16

9.
x	y
−1	1
2	−5
4	−9
6	−13
9	−19

10.
x	y
−4	2
−2	3
−1	3.5
2	5
6	7

Use the given rate of change and start value to complete each table.

11.
x	y
0	
1	
2	
3	
4	
5	

Rate of Change = +8
Start Value = 1

12.
x	y
−1	
	4.8
1	
2	
3	12.6
4	

Rate of Change = +2.6
Start Value = 4.8

13.
x	y
	18
−1	
0	
	3
3	
6	

Rate of Change = −5
Start Value = 8

14. Jim-Bob's Car Rental Company charges a set fee for renting a car and an additional amount per mile driven. Frank has rented from Jim-Bob's three times and his charges are shown in the table to the right.

Miles Driven	Cost
4	$17.60
10	$20.00
22	$24.80

    a. How much does Jim-Bob charge per mile driven?
    b. What is the set fee for renting a car at Jim-Bob's?
    c. How much would a car rental cost if Frank drove 30 miles? Show all work necessary to justify your answer.

15. Mark moved into a new house and believes his bedroom will soon be taken over by ants. In the table shown below, Mark records the number of ants in his bedroom on different days since he moved in.

Days Since Mark Moved In	Number of Ants
3	66
5	90
9	138
13	186
15	210

    a. How many ants are moving into Mark's bedroom each day?
    b. How many ants were in his room when he first moved in?
    c. If this pattern continues, how many ants will be in his room after 3 weeks? Use words and/or numbers to show how you determined your answer.

16. Mario rides his scooter to work each day. He is able to travel 0.5 miles per minute. He lives 4.7 miles from work.

Minutes Traveled	Distance to Work
0	4.7
1	
2	
3	

    a. Copy the table and fill in Mario's distance to work based on the number of minutes traveled. Continue the table until he has arrived at work.
    b. What is the rate of change in this situation?
    c. How long (to the nearest second) will it take for Mario to get to work? Explain how you arrived at your answer.

REVIEW

Each input-output table represents a real-world situation. Determine an appropriate range for the *y*-axis. State what increments you would use on the *y*-axis.

17.

Hours	Distance Traveled
0	45
1	75
2	105
3	135
4	165
5	195

18.

Days	Plant Height
0	0
1	0.2
2	0.4
3	0.6
4	0.8
5	1.0

19.

Lawns Mowed	Profit
0	−$50
1	−$30
2	−$10
3	$10
4	$30
5	$50

Tic-Tac-Toe ~ Challenging Tables

Find the equation that represents the linear relationship in each table. Show all work necessary to justify your answer.

1.

x	y
−3	1.5
0	2.7
2	3.5
6	5.1
8	5.9

2.

x	y
−8	1500
−3	875
4	0
9	−625
14	−1250

3.

x	y
−15	0
−12	1
−10	$1\frac{2}{3}$
−6	3
−1	$4\frac{2}{3}$

4.

x	y
−8	84
2	44
7	24
10	12
15	−8

5.

x	y
1	0.3
3	0.06
5	−0.18
7	−0.42
11	−0.9

6.

x	y
−2	2
3	$4\frac{1}{2}$
7	$6\frac{1}{2}$
10	8
12	9

Tic-Tac-Toe ~ Rate Applications

Rates of change are calculated in many situations. Determine the rate of change in each situation given below. Assume each situation forms a linear relationship. Use mathematics to justify each answer.

1. Ryan and Silas each bought a package of paper. Ryan bought 7 pencils with his paper for $1.96. Silas bought 12 of the same pencils with his paper for $3.36. Find the cost per pencil.

2. Kendra was at an elevation of −45 feet after 5 minutes of hiking. She was at −3 feet after eleven minutes. What was her rate of change in elevation in feet per minute?

3. Owen was 5.2 miles from home 10 minutes after school was over. He arrived home 30 minutes after school was over. What was his rate of speed going home from school?

Create a worksheet of 10 of your own rate problems. Type or neatly print the problems. Include the answers on a separate sheet of paper.

RECURSIVE ROUTINES TO EQUATIONS

LESSON 2.5

 Write linear equations from recursive routines.

In this lesson you will learn how to take a recursive routine and determine the **linear equation** that represents the situation. When the solutions of a linear equation are graphed, they form a line. Almost all linear equations are also linear functions. A **function** is a pairing of input and output values according to a specific rule. A function has exactly one output value for each input value.

One common form of a linear equation is $y = b + mx$ where b represents the start value and m represents the rate of change.

EXPLORE!

MODELING WITH EQUATIONS

Vendors at the local Farmer's Market sell a variety of produce and homemade products. Examine each vendor's situation and write a linear equation that models their profits.

Step 1: Peter sells corn at his booth in the Farmer's Market. His start-up cost for his business was $200 which he spent on seeds, fertilizer and water. What number would represent the start value for his situation: 200 or −200? Why?

Step 2: Peter earns $0.25 for each ear of corn he sells. What number represents his rate of change: 0.25 or −0.25? Why?

Step 3: One form of a linear equation is $y = b + mx$. The b represents the start value and m represents the rate of change. Write a linear equation to represent Peter's total profits if x represents the number of ears of corn sold and y is his total profit.

Step 4: Nakisha sells homemade candles at the Farmer's Market. She recorded her total profit for the first five candles sold in the table at the right. What is the start value for her business? What is the rate of change?

Step 5: Write a linear equation in the form of $y = b + mx$ to represent Nakisha's total profits at the Farmer's Market if x represents the number of candles sold and y is her total profit.

Candles Sold, x	Total Profits, y
0	−$10
1	−$7
2	−$4
3	−$1
4	$2
5	$5

Step 6: Nakisha sold a total of 24 candles. Use your linear equation to determine her total profit.

Lesson 2.5 ~ Recursive Routines to Equations

EXPLORE! (CONTINUED)

Step 7: Luke opened a booth at the Farmer's Market. He had a positive balance of $300 in his bank account. He paid $50 each week to rent his booth. He had trouble selling his products. His total savings is shown in the graph at the right. What is the recursive routine for this graph? Write a linear equation to represent this situation.

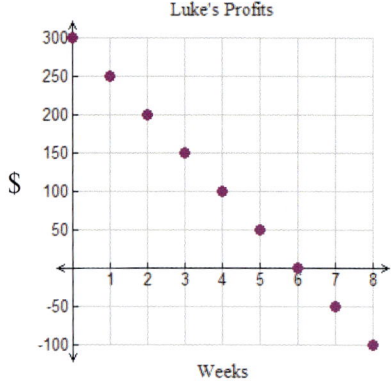

Step 8: One student summarized writing linear equations with the graphic below. How would you know whether to put a + or − between the start value and the rate of change in different situations?

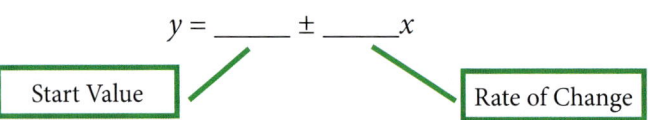

Step 9: Write a linear equation in the form $y = b + mx$ for each recursive routine:
    a. Start Value = 4   Rate of Change = +5
    b. Start Value = −7   Rate of Change = +2
    c. Start Value = 0   Rate of Change = −12
    d. Start Value = −8   Rate of Change = +0

WRITING LINEAR EQUATIONS FROM RECURSIVE ROUTINES

1. Determine the rate of change.
2. Determine the start value from a described situation, graph or table.
3. Write the linear equation by filling in the start value (*b*) and rate of change (*m*) in the equation:

$$y = b + mx$$

You have found the start value in a recursive routine by looking at tables. It is the number paired with an *x*-value of 0. The start value can also be located on a graph as the point on the *y*-axis.

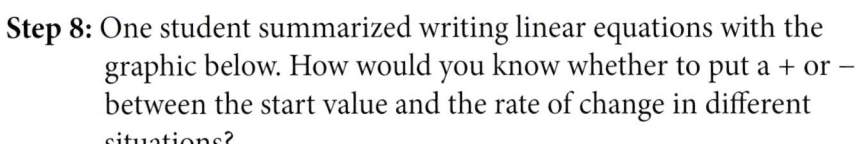

Start Value

x	y
−1	−5
0	−2
1	1
2	4
3	7

Linear equations represent the start value using the variable *b*. The start value (*b*) is also called the *y-intercept*. The *y*-intercept is the value of *y* where the graph crosses the *y*-axis or the number paired with an *x*-value of 0 in a table.

EXAMPLE 1

Write the linear equation for each recursive routine.
a. Rate of Change = +4
   Start Value = −23
b. Rate of Change = −0.3
   y-intercept = 2.8

c. Rate of Change = +2
   y-intercept = 0
d. Rate of Change = 0
   Start Value = 7

SOLUTIONS

a. Insert the rate of change and the start value into the equation $y = b + mx$. Remember, the rate of change is always the coefficient of the x-variable.
$$y = -23 + 4x$$

b. Insert the rate of change and the y-intercept into the equation. Remember, the y-intercept is just another way to refer to the start value.
$$y = 2.8 - 0.3x$$

c. Insert the rate of change and the y-intercept into the equation.
$$y = 0 + 2x \rightarrow y = 2x$$

> Adding 0 does not affect the value of the equation. It does not need to be written.

d. A rate of change equal to 0 cancels out the x-term.
$$y = 7 + 0x \rightarrow y = 7$$

EXAMPLE 2

Determine the rate of change and y-intercept (start value) for each table. Write a linear equation that represents each table.

a.
x	y
0	6
1	4
2	2
3	0
4	−2
5	−4

b.
x	y
−1	−2
1	8
4	23
6	33
9	48
12	63

SOLUTIONS

a. The start value is the y-value that is paired with the x-value of 0. The start value is 6.

The rate of change is calculated by determining the change in the y-values divided by the change in the x-values.

x	y
0	6
1	4
2	2
3	0
4	−2
5	−4

Change in x-values = +1   Change in y-values = −2

$$\text{Rate of Change} = \frac{\text{Change in } y\text{-values}}{\text{Change in } x\text{-values}} = \frac{-2}{+1} = -2$$

Linear Equation: $y = 6 - 2x$

Lesson 2.5 ~ Recursive Routines to Equations

EXAMPLE 2
SOLUTIONS
(CONTINUED)

b. Calculate the rate of change first when the start value is not given in the table.

x	y
−1	−2
1	8
4	23
6	33
9	48
12	63

Change in x-values = +2   Change in y-values = +10

$$\text{Rate of Change} = \frac{\text{Change in } y\text{-values}}{\text{Change in } x\text{-values}} = \frac{+10}{+2} = +5$$

To find the start value, use the rate of change to find the y-value that is paired with the x-value of 0.

x	y
−1	−2
0	3

+5

A rate of change of +5 means the y-value increases by 5 each time the x-value increases by 1. The x- and y-values for the step before zero are given. Add 5 once to the y-value to get the start value: −2 + 5 = 3.

Linear Equation: $y = 3 + 5x$

EXERCISES

Write the equation for each recursive routine.

1. Rate of Change = +8
Start Value = −6

2. Rate of Change = $-\frac{1}{2}$
y-intercept = $3\frac{1}{4}$

3. Rate of Change = +7.1
Start Value = 0

4. Rate of Change = −3
y-intercept = 7

5. Start Value = −10
Rate of Change = 0

6. y-intercept = 120
Rate of Change = −54

Determine the rate of change and y-intercept for each table. Write a linear equation that represents each table.

7.

x	y
0	4
1	12
2	20
3	28
4	36

8.

x	y
0	12
1	11
2	10
3	9
4	8

9.

x	y
−2	2
−1	4
0	6
1	8
3	12

Determine the rate of change and *y*-intercept for each table. Write a linear equation that represents each table.

10.
x	y
−1	29.5
0	31
3	35.5
4	37
7	41.5

11.
x	y
−2	−1
2	15
5	27
7	35
10	47

12.
x	y
4	1
6	2
10	4
13	5.5
18	8

13. Danna was able to finish 12 of her math homework problems at school. At home, she can do 4 problems every 2 minutes.
 a. How many problems does Danna complete each minute?
 b. What is Danna's start value for her homework on this particular day?
 c. Write a linear equation that represents this situation.
 d. What do the *x*-values represent in this equation?
 e. What do the *y*-values represent in this equation?

14. Jermaine wants to write a linear equation to help him calculate how much money he has saved based on the number of days he has been saving. He begins with nothing in his savings. He saves $6 per day.
 a. What is the linear equation that represents this situation?
 b. How much will he have saved after 12 days? Use words and/or numbers to show how you determined your answer.

15. Jack left a bottle of water sitting on the counter. When he first measured the temperature, it was 65° F. Each hour, he measured the temperature. It remained at 65° F.
 a. What is the *y*-intercept in this situation?
 b. What is the rate of change?
 c. Write a linear equation to represent the water's temperature based on the number of hours that have passed.

16. Shannon climbed to the top of a very tall slide and sent a ball down the slide. The top of the slide is 32 feet off the ground. The ball took only 4 seconds to make it to the bottom of the slide.
 a. What linear equation represents the ball's height off the ground based on the number of seconds it has traveled?
 b. How high off the ground was the ball after 2.2 seconds? Explain how you know your answer is correct.

Copy each table. Determine the rate of change and *y*-intercept. Fill in the missing values and write the linear equation that represents the table.

17.
x	y
0	−1
1	2
2	5
3	
4	

18.
x	y
−1	24
0	23.5
1	23
2	
3	

19.
x	y
0	14
1	25
2	36
5	
7	

Lesson 2.5 ~ Recursive Routines to Equations

REVIEW

Copy and complete each table by evaluating the given expression for the values listed.

20.

x	0.5x − 1	Output
−1		
0		
1		
2		
4		

21.

x	−2x + 7	Output
−2		
−1		
0		
3		
5		

Determine the rate of change for each situation.

22. $72 in 8 hours

23. 140 in 4 hours

24. 963 words in 9 minutes

25. 50 points for 4 assignments

Tic-Tac-Toe ~ Geometric Sequences

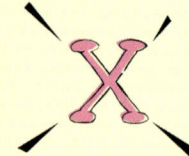

A geometric sequence is a list of numbers created by multiplying the previous term in the sequence by a common ratio. A geometric sequence can be described by a recursive routine with a start value and an operation. For example:

Start Value = 4   Operation = × 2   Sequence: 4, 8, 16, 32, 64, …

Since geometric sequences are not linear, they will not be represented by the equation $y = b + mx$. Geometric sequences can be represented by the equation $y = b \cdot m^x$ where b is the start value and m is the amount used for the repeated multiplication. For the example above, an equation to represent this sequence is $y = 4 \cdot 2^x$.

Copy and complete each geometric sequence. Give the start value and the operation.

1. 5, 15, _____, 135, _____

2. 2, −4, _____, −16, 32, _____

3. 120, _____, 30, 15, _____

4. −100, 10, _____, _____, _____

5. Find the next three terms in each geometric sequence in **Exercises #1 through #4**.

6. Write an equation to represent each geometric sequence in #1-4. Use your equation to find the 12th term.

7. Create two of your own geometric sequences. Record the start value, operation and the first eight terms. Write an equation representing each sequence.

Lesson 2.5 ~ Recursive Routines to Equations

INPUT-OUTPUT TABLES FROM EQUATIONS

LESSON 2.6

 Determine the rate of change and start value from linear equations.
Create input-output tables from linear equations.

The Commutative Property of Addition states that numbers can be added in any order. This can be applied in a linear equation. In **Lesson 2.5**, linear equations were shown in the form $y = b + mx$. Based on the Commutative Property, this equation can also be written $y = mx + b$. For example:

$$y = 7 + 2x \rightarrow y = 2x + 7$$
$$y = 4 - 3x \rightarrow y = -3x + 4$$

Notice that the "−" belongs to the rate of change and must move with it.

EXPLORE!

Use the linear equations given in the box to complete this activity.

Step 1: List the equation(s) that have a negative rate of change.

Step 2: List the equations(s) that have a positive y-intercept.

Step 3: Which equation has a start value of zero? How do you know?

Step 4: Which equation has a rate of change equal to zero? How do you know?

Step 5: Create your own linear equation that fits the given description.
  a. A positive start value and negative rate of change.
  b. A rate of change equal to zero.
  c. A start value equal to zero.
  d. A negative rate of change and a negative y-intercept.

LINEAR QUALITIES

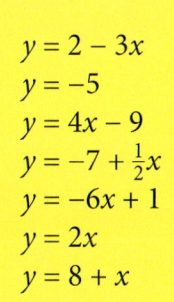

$y = 2 - 3x$
$y = -5$
$y = 4x - 9$
$y = -7 + \frac{1}{2}x$
$y = -6x + 1$
$y = 2x$
$y = 8 + x$

There are times when you are given the equation that describes a relationship between two pieces of information. An equation is useful in creating other ways to display the information. The most common ways of displaying data are through graphs, tables and words.

Over summer break Josie went to the mall with her friends. At noon, they left the mall and began walking. The equation $y = 2 + 3x$ represents Josie's distance (y) from her home. The x represents the number of hours she has walked.

Josie walked for at least three hours. An input-output table of values can be created that shows how far Josie is from home based on how long she walked. This can be done by identifying the *y*-intercept and rate of change of the situation.

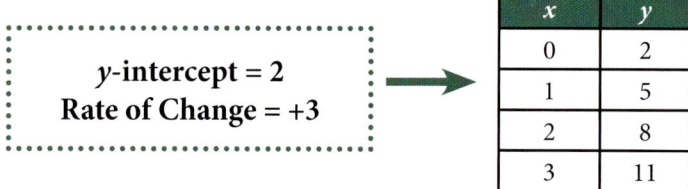

Graphs are another way to visually display equations. Take each ordered pair from the table and graph it on a coordinate plane. Since Josie is continually walking, the points can be connected to form a straight line.

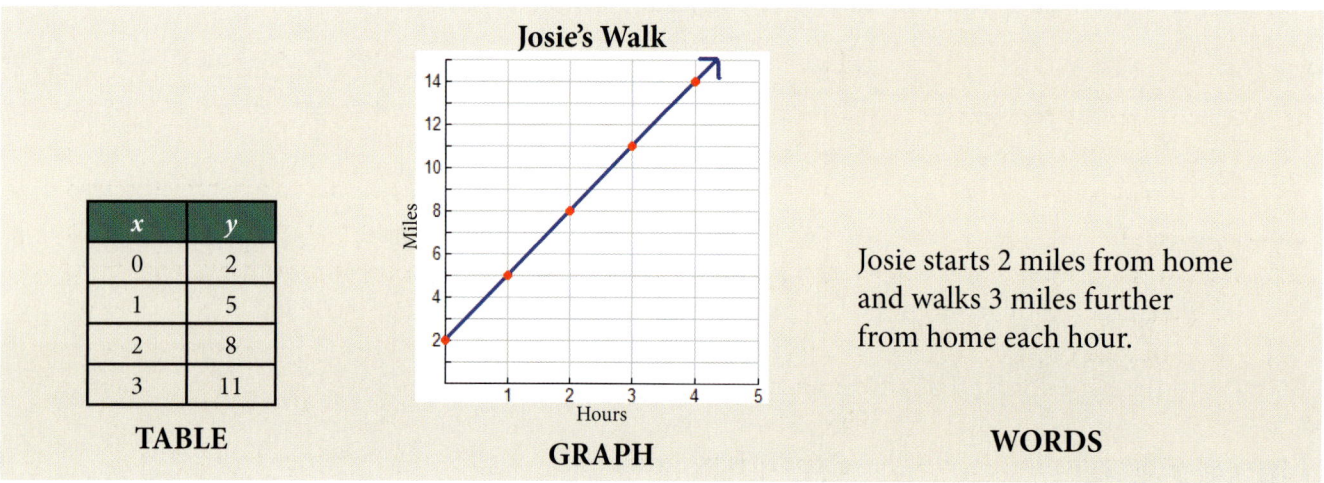

Josie starts 2 miles from home and walks 3 miles further from home each hour.

TABLE　　GRAPH　　WORDS

Input-output tables can be completed by evaluating the equation for the values given in the table.

EXAMPLE 1　Use the linear equation to complete the input-output table.　$y = 3x + 8$

x	y
−4	
7	
16	
29	

SOLUTION　Substitute each *x*-value into the equation to determine the *y*-values.

x	3x + 8	y
−4	3(−4) + 8	−4
7	3(7) + 8	29
16	3(16) + 8	56
29	3(29) + 8	95

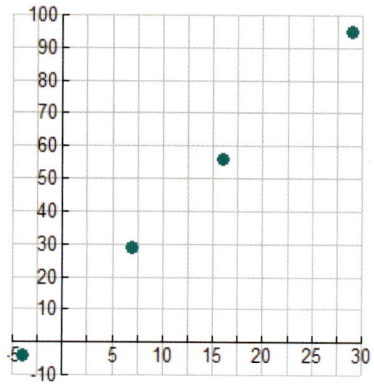

Any input-output table can be turned into ordered pairs and graphed. When dealing with linear equations, graphing is a great way to double-check the calculations as all points should be in a straight line.

EXERCISES

Determine the rate of change and the *y*-intercept from the given equations.

1. $y = 8 + 2x$
2. $y = 3x - 11$
3. $y = x - 4$
4. $y = 5 - 4x$
5. $y = -\frac{1}{4}x$
6. $y = -1$
7. $y = \frac{2}{3}x - 8$
8. $y = 6$
9. $y = 2 - \frac{4}{7}x$

Given the equation, copy and complete the input-output tables.

10. $y = 2x - 3$

x	2x − 3	y
0		
3		
9		
10		
13		

11. $y = x + 9$

x	x + 9	y
−4		
0		
2		
5		
21		

12. $y = -10 + 6x$

x	−10 + 6x	y
−7		
−3		
1		
4		
15		

13. $y = -3x$

x	y
−3	
−1	
6	
10	
20	

14. $y = \frac{1}{2}x + 1$

x	y
−6	
−1	
4	
6	
11	

15. $y = 5$

x	y
−4	
−3	
0	
1	
5	

16. Gracie planted a marigold in June. She measured its height each week and found that the height of the plant could be represented by the equation $y = 3 + 0.5x$ where *x* represents the number of weeks that have passed and *y* represents the height of the plant in inches.
    a. Copy and complete the table to show the height of the marigold through the summer.
    b. Graph the ordered pairs on a coordinate plane.
    c. The flower stopped growing when it reached 1 foot tall. How many weeks had passed since the flower was planted? Show all work necessary to justify your answer.

x	y
0	
4	
7	
10	
12	

17. Debra graphed the points from the table at the right. The ordered pairs she graphed were (8, 2), (10, 3) and (12, 4). Explain her mistake to her and fix it by writing the correct ordered pairs.

x	y
2	8
3	10
4	12

18. Nguyen argued that an equation is much more useful than a table or graph. Give an example of a situation where a graph may be more useful to someone than just the equation.

Lesson 2.6 ~ Input-Output Tables from Equations

19. Star is able to run 6.8 meters per second when she is sprinting. She wants to figure out how many meters (y) she can run based on the number of seconds (x) she has run. She developed an equation to help her: $y = 6.8x$.

x seconds	y meters run
10	
25	
40	
60	
100	

    a. Copy and complete the table using Star's equation.
    b. Four hundred meters is approximately a quarter of a mile. About how long would it take Star to run one-quarter of a mile? Is this reasonable? Why or why not?
    c. Star decides she is going to run for one hour. How many seconds is this?
    d. According to her equation, how many meters would she run in one hour?
    e. There are about 1,600 meters in a mile. Convert your answer from **part d** into miles.
    f. Is this answer reasonable? Why or why not?

20. During the summer, Jorge works at a kids camp. He was given $100 for signing on for the summer and then is paid an additional $35 per day of work. Copy and complete the table that shows Jorge's total earnings based on how many days he works.

x days	y earnings
6	
20	
32	
44	
50	

REVIEW

Solve each equation. Show all work necessary to justify your answer.

21. $2x + 7 = 22$

22. $\frac{x}{5} - 9 = -3$

23. $-x + 2 = 8$

24. $4(x + 7) = 12$

25. $6x + 1 = 5x + 4$

26. $2(3x - 2) = 38$

27. $2x - 5 = 5x + 28$

28. $6 + \frac{x}{3} = 2$

29. $8 = 23 - 5x$

Tic-Tac-Toe ~ Writing Equations From Tables

Creating equations from input-output tables is a difficult process. Create a worksheet that steps a student through the process of finding the rate of change, the start value and then writing the equation. Include tables that have a start value listed in the table and some that do not. Turn in a blank copy of the worksheet and an answer key.

CALCULATING SLOPE FROM GRAPHS

LESSON 2.7

 Use slope triangles to find the slope of lines.

Up to this point you have looked at linear relationships by examining their rates of change and start values, which are also called *y*-intercepts. The rate of change tells you how much the *y*-value should increase or decrease as the independent variable (*x*) increases.

Rate of change is also known as **slope**. Like rate of change, slope is used to describe the steepness of a line. Slope is the ratio of the vertical change (the rise) to the horizontal change (the run). The easiest way to calculate the slope of a line when it is graphed is to create a **slope triangle**. A slope triangle is formed by drawing a right triangle where one leg of the triangle represents the vertical rise and the other leg is the horizontal run. The hypotenuse of the triangle (the longest side) is part of the line itself. Start at the point furthest to the left and go up or down to draw your first leg. Then draw your second leg to the right.

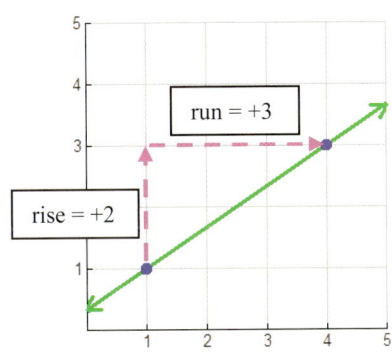

$$\text{Slope} = \frac{\text{rise}}{\text{run}} = \frac{+2}{+3} = \frac{2}{3}$$

A line can have a slope that is positive, negative, zero or undefined.

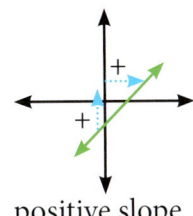
positive slope

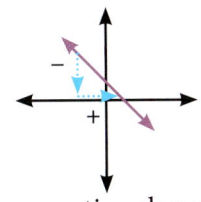
negative slope

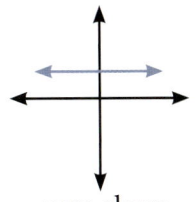
zero slope

undefined slope

FINDING SLOPE

The slope of a line is the ratio of the change in *y*-values to the change in *x*-values.

$$\text{Slope} = \text{Rate of Change} = \frac{\text{Change in } y\text{-values}}{\text{Change in } x\text{-values}} = \frac{\text{rise}}{\text{run}}$$

EXAMPLE 1 Draw a slope triangle for each line (when possible) and identify the slope of the line.

a.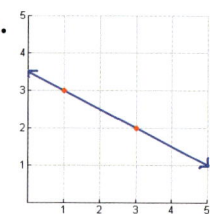
b.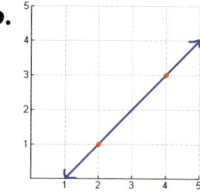
c.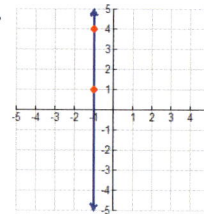
d.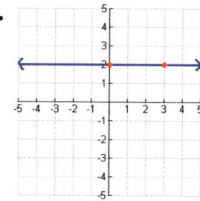

SOLUTIONS

a.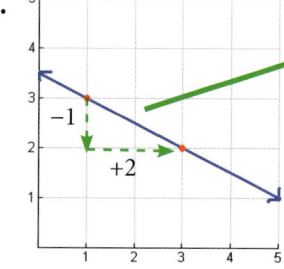

$$\text{Slope} = \frac{\text{rise}}{\text{run}} = \frac{-1}{+2} = -\frac{1}{2}$$

b.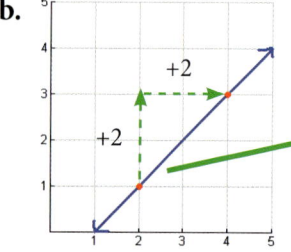

$$\text{Slope} = \frac{\text{rise}}{\text{run}} = \frac{+2}{+2} = 1$$

c.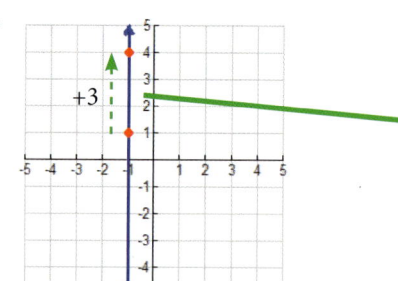

$$\text{Slope} = \frac{\text{rise}}{\text{run}} = \frac{+3}{0} = \text{undefined}$$

d.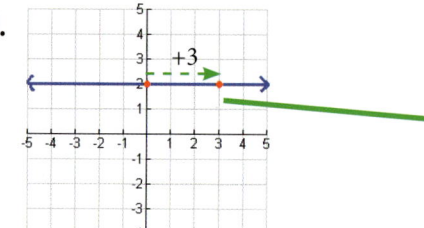

$$\text{Slope} = \frac{\text{rise}}{\text{run}} = \frac{0}{+3} = 0$$

There are times when a graph is not provided. You may only be given a table of values or two ordered pairs. In each of these situations, you can graph the points and then create a slope triangle to calculate the slope.

EXAMPLE 2 Graph the line that goes through the given points, draw a slope triangle and give the slope.

a.
x	y
2	7
5	1

b. (0, 1) and (2, 4)

SOLUTIONS

a. Graph the points.

Draw the slope triangle. Start at the point furthest to the left.

Determine the lengths of the legs of the triangle.
Slope = $\frac{\text{rise}}{\text{run}} = \frac{-6}{+3} = -2$

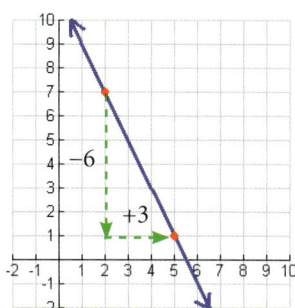

b. Graph the points.

Draw the slope triangle. Start at the point furthest to the left.

Determine the lengths of the legs of the triangle.
Slope = $\frac{\text{rise}}{\text{run}} = \frac{+3}{+2} = \frac{3}{2}$

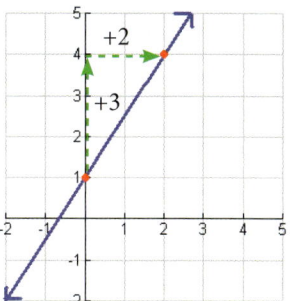

When given the slope of a linear relationship you can draw a line that has the given slope using rise over run. When a slope is written without a denominator, it is mathematically correct to place a 1 in the denominator. This does not change the value of the slope. For example:

$$3 \rightarrow \frac{3}{1}$$

EXAMPLE 3 Graph a line with a slope of 2.

SOLUTION No y-intercept was given so put a point anywhere on the graph.

The slope of 2 can be written as $\frac{2}{1}$.

From the point, rise +2 and run +1.

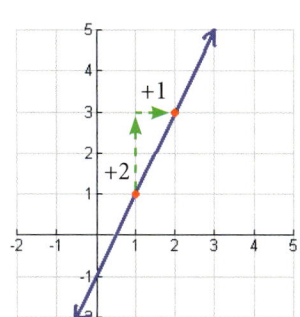

Lesson 2.7 ~ Calculating Slope from Graphs

EXERCISES

Find the slope of each line.

1.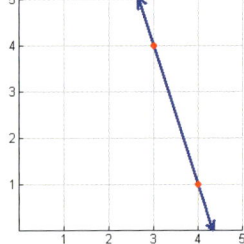
2.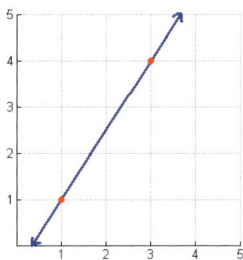
3.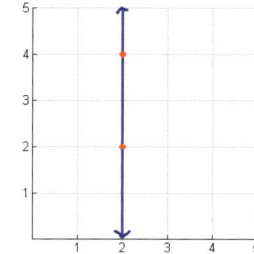
4.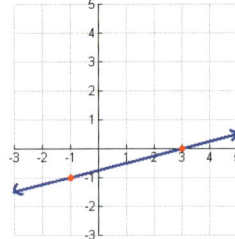
5.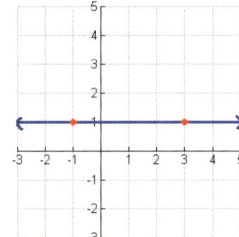
6.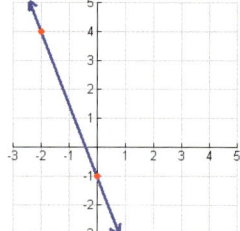

Use each table or graph to determine if the slope of the line is positive, negative, zero or undefined.

7.
x	y
0	12
1	9
2	6
3	3

8.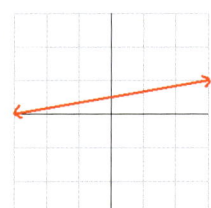

9.
x	y
4	7
7	7
12	7
16	7

10.

11.
x	y
−3	1
−1	3
3	7
6	10

12.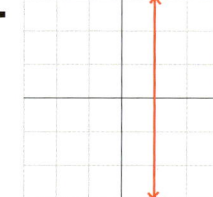

13. Karissa and Rider both calculated the slope of the same line. Rider says the slope is $\frac{-2}{3}$ and Karissa believes the slope is $\frac{2}{-3}$.
    a. On a coordinate plane, graph a line that goes through the origin and has a slope of $\frac{-2}{3}$.
    b. On a different coordinate plane, graph a line that goes through the origin and has a slope of $\frac{2}{-3}$.
    c. What do you notice about the two lines? The teacher says the slope is $-\frac{2}{3}$. Is this the same slope as found by Karissa, Rider or both? Explain your reasoning.

On a coordinate plane, graph a line with the given slope.

14. slope = $\frac{3}{4}$

15. slope = −3

16. slope = 0

Draw a line through the given point that has the given slope on a coordinate plane. Name one other ordered pair that is on the line.

17. (1, 3), slope = −2

18. (−2, 1), slope = −$\frac{2}{5}$

19. (3, −4), slope = undefined

20. Barry is building a staircase from the first floor to the second floor. The height between the two floors is 12 feet. He wants the slope of the stairs to be $\frac{4}{3}$. What is the horizontal distance that the stairs will cover? Use mathematics to justify your answer.

21. A wheelchair ramp is being designed for the library entrance. The pavement pouring company advises that the slope of the ramp be $\frac{2}{7}$. If the entrance to the library is 6 feet above ground, how long will the ramp need to be? Show all work necessary to justify your answer.

22. When finding the slope of a line on a graph, can you choose any two points on the line? Support the answer by determining the slope of the line shown at the right using three different slope triangles.

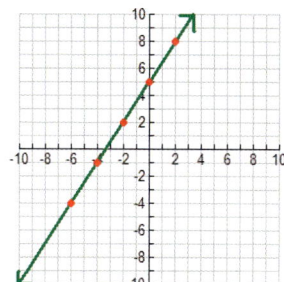

REVIEW

Evaluate each expression using the order of operations. Write all answers in simplest form.

23. $\frac{7-3}{12-6}$

24. $\frac{-1-8}{6-3}$

25. $\frac{4-(-5)}{4-3}$

26. $\frac{8-8}{1-7}$

27. $\frac{7-2}{5-5}$

28. $\frac{-3-(-2)}{7-(-10)}$

Tic-Tac-Toe ~ Children's Story

Sequences occur in many real-world situations. Create a children's book that incorporates the concept of recursive sequences and recursive routines. The character(s) in your book should encounter sequences in a variety of real-world situations. The plot should include the character(s) needing to find the start values, operations and specific term in recursive sequences. Your book should have a cover, illustrations and a story line that is appropriate for children.

THE SLOPE FORMULA

LESSON 2.8

 Find the slope of a line using the slope formula.

EXPLORE! **FIND THAT FORMULA**

Ginger got a job in downtown Oklahoma City. She bought a parking pass at a garage not far from her place of work. The table shows her total parking expenses based on the number of weeks she has been parking at the garage.

Week x	Total Expenses, y
6	$50
10	$74
12	$86
24	$158

Step 1: Calculate the rate of change (the change in y over the change in x) for the table above.

Step 2: Graph the ordered pairs on a Quadrant I coordinate plane like the one shown below. Draw a line through the points.

Step 3: Make a slope triangle and determine the slope of the line.

Step 4: What do you notice about the rate of change and the slope of the line?

Step 5: If you were given the table of values in the table at right, what would the rate of change (or slope) ratio look like?

x	y
x_1	y_1
x_2	y_2

Step 6: The ratio developed in **Step 5** is called the "Slope Formula". The subscripts identify two different points. Try your formula on these points from the table above: (6, 50) and (12, 86). Did you get the same slope as you did in **Steps 1 and 3**?

Step 7: You have learned three methods for finding slope: rate of change, slope triangles and the slope formula. Which method do you like the best? Why?

THE SLOPE FORMULA

The formula for the slope of a line that goes through a point with coordinates (x_1, y_1) and another point with coordinates (x_2, y_2) is

$$\text{Slope} = \frac{y_2 - y_1}{x_2 - x_1}$$

Subscripts designate which point you are using in your calculation. You read x_1 as "x sub one". Think of it as saying "the x-coordinate of the first point".

EXAMPLE 1 Use the slope formula to find the slope of each line that passes through the given points.
a. (3, 2) and (8, 5)
b. (1, −1) and (3, −5)
c. (6, −2) and (6, 4)

Sometimes it helps to write the subscript letters over your points to stay organized:

$$\begin{array}{c} x_1 \ y_1 \\ (3, 2) \end{array}$$

SOLUTIONS

a. Let (3, 2) be (x_1, y_1) and (8, 5) be (x_2, y_2).

$$\frac{y_2 - y_1}{x_2 - x_1}$$

Substitute the numbers into the slope formula. $\dfrac{5 - 2}{8 - 3} = \dfrac{3}{5}$

b. Let (1, −1) be (x_1, y_1) and (3, −5) be (x_2, y_2).

$$\frac{y_2 - y_1}{x_2 - x_1}$$

Substitute the numbers into the slope formula. $\dfrac{-5 - (-1)}{3 - 1} = \dfrac{-4}{2} = -2$

c. Let (6, −2) be (x_1, y_1) and (6, 4) be (x_2, y_2).

$$\frac{y_2 - y_1}{x_2 - x_1}$$

Substitute the numbers into the slope formula. $\dfrac{4 - (-2)}{6 - 6} = \dfrac{6}{0} =$ undefined

It is impossible to divide by 0 so the slope is undefined.

You have learned three methods for calculating slope. All three methods will work in any situation. Depending on the way the linear relationship is presented, there may be one method that is easier to use than the other two methods.

RATE OF CHANGE   Easiest method when information is presented in an input-output table.

SLOPE TRIANGLE   Easiest method when information is presented in a graph.

SLOPE FORMULA   Easiest method when given two ordered pairs.

EXERCISES

Find the slope of the line that passes through the given points.

1. (4, 7) and (6, 10)

2. (0, 8) and (3, 5)

3. (−1, 11) and (4, 11)

4. (−6, −1) and (2, 0)

5. (7, −2) and (7, 9)

6. (6, 4) and (2, 10)

7. (−2, 10) and (3, 10)

8. (0, −4) and (5, 0)

9. (9, 8) and (4, 18)

10. In each part below, explain what method you would choose to calculate the slope. Then find the slope.

a.
x	y
2	2
5	11
9	23
14	38

b. A line through (1, 5) and (−2, 9)

c.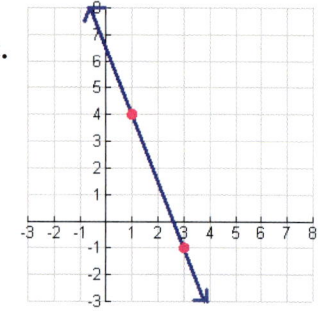

11. Consider the line through the points (3, 8) and (8, 12).
  a. Find the slope.
  b. Convert your slope fraction to a decimal. Remember that this is your rate of change.
  c. The rate of change tells you how much to increase or decrease for every one-unit step. Copy and complete the table using the rate of change.
  d. How many ordered pairs for this line do you have now? How many more could you figure out?

x	y
3	8
4	
5	
6	
7	
8	12

12. Tanika joined a gym. At 4 months, she had paid a total of $94 in membership fees. After 9 months, she had paid a total of $204 in membership fees. Let x represent the number of months she has been a member and y represent the total she has paid in membership fees.
  a. Write two ordered pairs to represent Tanika's gym membership information.
  b. Find the slope of the line that contains the two points in **part a**.
  c. What does the slope represent in terms of Tanika's gym membership fee?

13. Scott skis at Clear Creek Pass during the winter. His favorite run has a vertical descent of 1,560 feet. The run covers a horizontal distance of 3,900 feet. What is the slope of this particular run? Give the answer as a fraction and as a decimal.

14. Devin drove up a hill. After he drove 4 minutes, he was at an elevation of 620 feet. After he drove 12 minutes he was 1,580 feet high.
  a. Should the x-values represent minutes or feet? Why?
  b. Calculate the rate of change in this situation. What method did you use and why?
  c. If he continues to climb at this rate, how much elevation will he gain in the next 15 minutes? Use mathematics to justify your answer.

84 Lesson 2.8 ~ The Slope Formula

15. Nigel put a two-liter bottle of soda in his locker. He did not realize there was a hole in the bottom of the container and that the liquid had slowly dripped out. The graph represents the amount of soda left in the bottle based on the amount of time that had passed.

    a. What is the slope of the line?
    b. What does the value of the slope represent in this situation?
    c. How many hours had passed when Nigel's soda bottle became empty? Explain how you know.

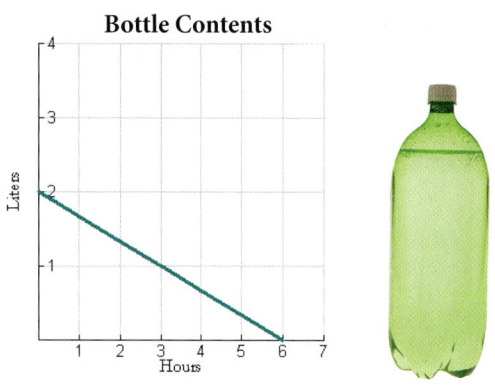

16. Why do you think the term "rate of change" is used more often than the word "slope" in real-world applications?

17. Create a set of three ordered pairs that fall on a line with a slope of $-\frac{3}{4}$. Explain how you know your answer is correct.

REVIEW

Match each recursive rule with its linear equation.

18. Start Value: 4   Slope: $\frac{1}{2}$

19. Start Value: −2   Rate of Change: $+\frac{1}{2}$

20. y-intercept: 1   Slope: 3

21. Start Value: $\frac{1}{2}$ Rate of Change: −2

22. y-intercept: 3   Rate of Change: +1

23. Start Value: −2   Slope: 0

24. y-intercept: 0   Rate of Change: +2

A. $y = 3x + 1$

B. $y = \frac{1}{2} - 2x$

C. $y = 2x$

D. $y = x + 3$

E. $y = 4 + \frac{1}{2}x$

F. $y = \frac{1}{2}x - 2$

G. $y = -2$

Tic-Tac-Toe ~ Slope Methods

You have learned to find the slope of a line when given either a graph, table or two ordered pairs. Create a flip book that explains how to use the slope formula, slope triangles and input-output tables to find slope. Include examples and diagrams.

REVIEW

BLOCK 2

Vocabulary

axes
coordinate plane
dependent variable
function
independent variable
linear equation

ordered pair
origin
quadrants
rate of change
recursive routine

recursive sequence
slope
slope triangle
x-axis
y-axis
y-intercept

Write recursive routines and create recursive sequences.
Create linear plots for recursive sequences.
Represent recursive routine applications with graphs, tables and words.
Calculate rates of change and start values.
Write linear equations from recursive routines.
Determine the rate of change and start value from linear equations.
Create input-output tables from linear equations.
Use slope triangles to find the slope of lines.
Find the slope of a line using the slope formula.

Lesson 2.1 ~ Recursive Routines

Copy each sequence of numbers and fill in the missing values. Identify the start value and the operation that must be performed to arrive at the next term.

1. 8, 1, –6, ____, ____, ____

2. 92, 110, ____, 146, ____, ____

3. 7, 7.6, ____, ____, 9.4, ____

4. ____, ____, 19, 22, 25, ____

For each sequence below, describe the recursive routine (start value and operation) and give the next three terms in the sequence.

5. 18, 5, –8, –21, …

6. $\frac{2}{5}$, $\frac{4}{5}$, $1\frac{1}{5}$, $1\frac{3}{5}$, …

7. Draw the next two figures in the following pattern. Each block is 1 unit by 1 unit.
   a. What is the perimeter of the first figure?
   b. What is the perimeter of the second figure? The third figure?
   c. Write the recursive routine (start value and operation) that describes the perimeters.
   d. What is the perimeter of the seventh figure in this pattern?
      Use words and/or numbers to show how you determined your answer.

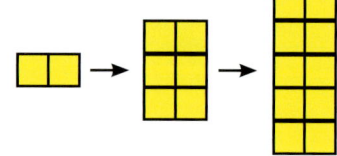

Lesson 2.2 ~ Linear Plots

Describe the linear relationship given by the *y*-coordinates on each linear plot below by stating the start value and operation. Create an input-output table showing the ordered pairs represented by each plot.

8.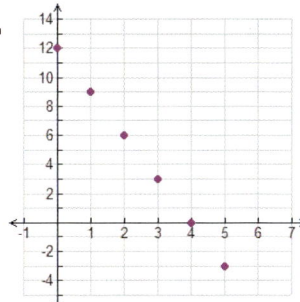

9.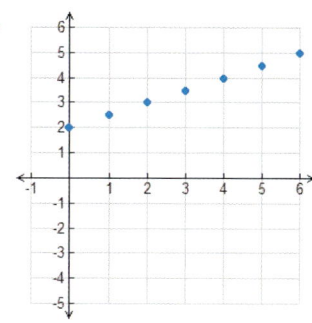

10. Create a linear plot for the first five ordered pairs in the given recursive routine describing the *y*-coordinates.
    a. Start Value: 11
       Operation: Subtract 2
    b. Start Value: −4
       Operation: Add $\frac{1}{2}$

Lesson 2.3 ~ Recursive Routine Applications

11. Determine an appropriate range for the *y*-axis and state what increments should be used on the graph.

    a.
Hours	Miles Traveled
0	55
1	110
2	165
3	220
4	275

    b.
Years	Savings
0	$10,200
1	$9,400
2	$8,600
3	$7,800
4	$7,000

12. LaQuisha borrowed $150 from her parents to buy school clothes. Each week she uses $12 of her allowance to repay her parents.
    a. Write a recursive routine that describes LaQuisha's balance based on the number of weeks that have passed since she borrowed the money.
    b. Fill in an input-ouput table that will give her balance for 0 to 5 weeks.
    c. Create a linear plot that shows LaQuisha's balance through the first five weeks.
    d. How many weeks will it take before LaQuisha has repaid her parents?
    e. How much will her last payment be? Show all work necessary to justify your answer.

13. In Oregon, the cost of a gallon of gasoline in January 2013 averaged $3.35. One analyst predicted that the cost of gas would rise $0.40 per year.
    a. Write a recursive routine that describes the price of gasoline based on the number of years that have passed since January 2013.
    b. Create an input-output table for the value of a gallon of gas for 0 to 5 years. Let 2013 represent year 0.
    c. Create a linear plot that shows the cost of a gallon of gasoline through the first five years.
    d. Determine how many years it will take before the cost of gasoline will be over $10 per gallon according to this analyst's prediction. Use words and/or numbers to show how you determined your answer.

Block 2 ~ Review

Lesson 2.4 ~ Rate of Change

Determine the rate of change for each situation.

14. Jordan collected 28 coins in 4 days.

15. Rebecca drove 270 miles in 6 hours.

16. Shea spent $4.56 for 12 doughnuts.

Determine the rate of change and start value for each table.

17.

x	y
0	2
1	9
2	16
3	23
4	30

18.

x	y
0	33
3	0
5	−22
6	−33
8	−55

19.

x	y
−1	9
1	17
3	25
5	33
8	45

20. Maria began an exercise plan at the beginning of the school year. In the table shown below, Maria records her weight at different points during the school year. Assume Maria loses the same amount every week.
  a. How many pounds is Maria losing each week?
  b. How much did she weigh when she first started this exercise plan?
  c. If this pattern continues, how much will she weigh 17 weeks into her exercise plan?
  d. Does it make sense that this pattern will continue throughout the whole school year? Why or why not?

Weeks Since Maria Started Exercising	Her Weight
1	161
5	153
7	149
10	143

Lesson 2.5 ~ Recursive Routines to Equations

Write the linear equation for each recursive rule.

21. Rate of Change = +6
Start Value = −2

22. Rate of Change = $-\frac{1}{2}$
y-intercept = 6

23. Rate of Change = +3.8
Start Value = 1

Determine the rate of change and y-intercept for each table. Write a linear equation that represents each table.

24.

x	y
0	39
1	33
2	27
3	21
4	15

25.

x	y
−2	2.9
−1	6.4
0	9.9
1	13.4
3	20.4

26.

x	y
−2	−15
2	1
5	13
7	21
10	33

27. Madison receives $42 at the beginning of the month to use for lunch money. Each day at school she buys the lunch special and a milk. This costs her $3.10. Madison wants to develop a linear equation to calculate how much money she has left based on the number of days she has bought lunch during a month.

    a. What is the *y*-intercept in this situation?
    b. What is the rate of change?
    c. Write a linear equation to represent the amount of lunch money Madison has left based on the number of days this month she has bought lunch.
    d. How much money will she have left after 5 days of buying lunch? Show all work necessary to justify your answer.

Lesson 2.6 ~ Input-Output Tables from Equations

Determine the rate of change and the *y*-intercept from the given equations.

28. $y = 5x + 1$

29. $y = 7 - 6x$

30. $y = 2 + \frac{1}{2}x$

31. $y = 2x + 7$

32. $y = \frac{2}{9}x$

33. $y = 8$

Given the equation, copy and complete the input-output tables.

34. $y = 3x + 1$

x	y
0	
3	
9	
10	
13	

35. $y = \frac{1}{3}x + 4$

x	y
0	
3	
4	
6	
11	

36. $y = 8x + 7$

x	y
−7	
−3	
1	
4	
15	

Lesson 2.7 ~ Calculating Slope from Graphs

Find the slope of each line.

37.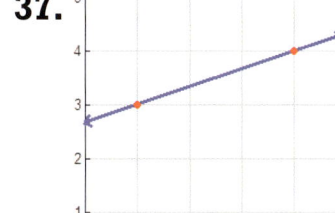

38.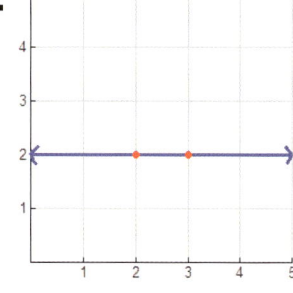

39.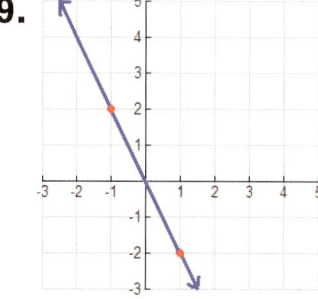

Create a coordinate plane and draw a line through the given point that has the given slope. Name one other ordered pair that is on the line.

40. (0, 0), slope = $\frac{1}{2}$

41. (−1, 3), slope = −4

42. (0, 2), slope = 0

43. A ladder leaned against a wall at a slope of $\frac{4}{3}$. The top of the ladder was 12 feet off the ground. How far is the bottom of the ladder from the base of the wall? Use mathematics to justify your answer.

44. The ladder in **Exercise 43** is leaning against another wall only 10 feet off the ground. The bottom of the ladder is 10 feet away from the base of the building.
   a. What is the slope of the ladder in this position?
   b. Is it steeper than the slope of the ladder in **Exercise 43**? Explain your reasoning.

Lesson 2.8 ~ The Slope Formula

Find the slope of the line that passes through the given points.

45. (3, 2) and (6, 6)

46. (0, 5) and (3, 8)

47. (−2, 9) and (8, 9)

48. (3, −2) and (5, −5)

49. (3, 2) and (3, 4)

50. (7, 5) and (3, 11)

51. For each part below, explain which method you would choose to calculate the slope. Then find the slope.

a.
x	y
2	21
6	13
9	7
11	3

b. a line through (3, 6) and (−1, 11)

c.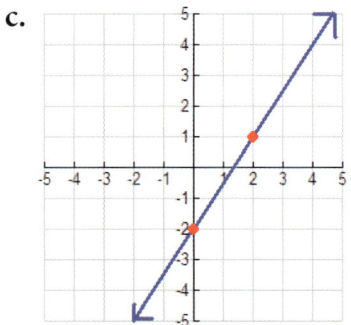

52. Angelina's family planned to rent a vacation home in the Poconos. For 3 nights it would cost a total of $510. For 5 nights it would cost a total of $800. Let *x* represent the number of nights the house would be rented and *y* represent the total cost.
   a. Write two ordered pairs to represent the house rental information.
   b. Find the slope of the line that contains the two points in **part a**.
   c. What does your slope represent in terms of Angelina's house rental?

53. Create a set of three ordered pairs that fall on a line with a slope of $\frac{1}{3}$. Explain how you know your answer is correct.

Tic-Tac-Toe ~ Similar Slope Triangles

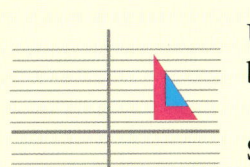

Use graph paper to draw three coordinate planes each from −10 to 10 on both axes.

Step 1: On the first coordinate plane, draw a line with a slope of $\frac{2}{3}$.

Step 2: Draw three different sizes of slope triangles on the line. For example, the graph below shows three different slope triangles for a line with a slope of $-\frac{1}{3}$.

Step 3: Write the slope fraction for each triangle.

Step 4: What do you notice about the slope fractions?

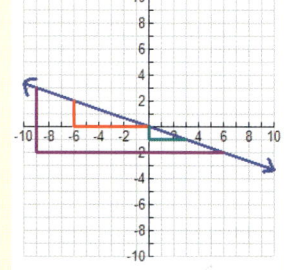

Step 5: On the second coordinate plane, draw a line with a slope of −2.
On the third coordinate plane, draw a line with a slope of $\frac{1}{4}$.
Repeat **Steps 2-4** for these two graphs.

Step 6: Can you draw any size of slope triangle on a line and get the correct slope? Are all slope triangles drawn on a given line similar? Summarize your results. Use math to support your answers.

CAREER FOCUS

KAREN
NEWSPAPER EDITOR

I am an editor for a daily newspaper. Our paper has a circulation of about 17,000. I coordinate local news coverage and oversee a staff of seven reporters. I also edit stories, decide when and where stories will run and design the local pages. Most of my time is spent working with the reporters on stories and laying out how each page will look. I also respond to telephone calls and emails from the public. Other duties include overseeing a monthly special section on homes and gardens and leading staff meetings twice a month.

Math is part of my job in at least a small way every day. I use ratios and percentages to figure out how articles and pictures of varying sizes fit into a given space. I approve time cards and expense reports so I need to double check the figures for correct addition and multiplication. Statistics are an important element of many news stories. I make sure numbers in the stories are correct and logical for the situation. For example, our paper may do a story on how crime has changed over the past three years. This might include overall figures, a breakdown of crime totals and percentages showing the increases or decreases from prior years. It might also compare types of crimes. The first thing I do is double check the figures to make sure the reported values correspond to the text. Secondly, if the story says overall crime is down but property crime is on the rise, the numbers we print had better show that fact.

I received a Master of Arts degree in journalism. I also have a bachelor's degree in Biology, though most people in my profession have degrees in journalism, writing or English. Salaries depend on where you work (the size of the city) as well as experience. A typical starting salary for a beginning copy editor is around $20,000 per year. A beginning city editor might start around $30,000 per year. An executive editor earns $50,000 per year or more.

Creating a newspaper is exciting. There is the pressure of deadlines, the challenge of editing, the creativity of page design and the sense of accomplishment that comes each day when you flip through the paper that you created. There is great satisfaction creating a product every day and knowing you did your best to make it meaningful for the thousands of people it reaches.

CORE FOCUS ON LINEAR EQUATIONS
BLOCK 3 ~ USING LINEAR EQUATIONS

LESSON 3.1	GRAPHING USING SLOPE-INTERCEPT FORM	95
LESSON 3.2	WRITING LINEAR EQUATIONS FOR GRAPHS	100
	EXPLORE! FIND THE EQUATION	
LESSON 3.3	WRITING LINEAR EQUATIONS FROM KEY INFORMATION	106
	EXPLORE! TRIANGLE LINES	
LESSON 3.4	DIFFERENT FORMS OF LINEAR EQUATIONS	111
	EXPLORE! ONE OF THESE THINGS	
LESSON 3.5	MORE GRAPHING LINEAR EQUATIONS	115
	EXPLORE! MATCH ME	
LESSON 3.6	GRAPHING LINEAR INEQUALITIES IN TWO VARIABLES	120
	EXPLORE! IN THE SHADE	
LESSON 3.7	INTRODUCTION TO NON-LINEAR FUNCTIONS	125
	EXPLORE! NON-LINEAR CURVES	
REVIEW	BLOCK 3 ~ USING LINEAR EQUATIONS	130

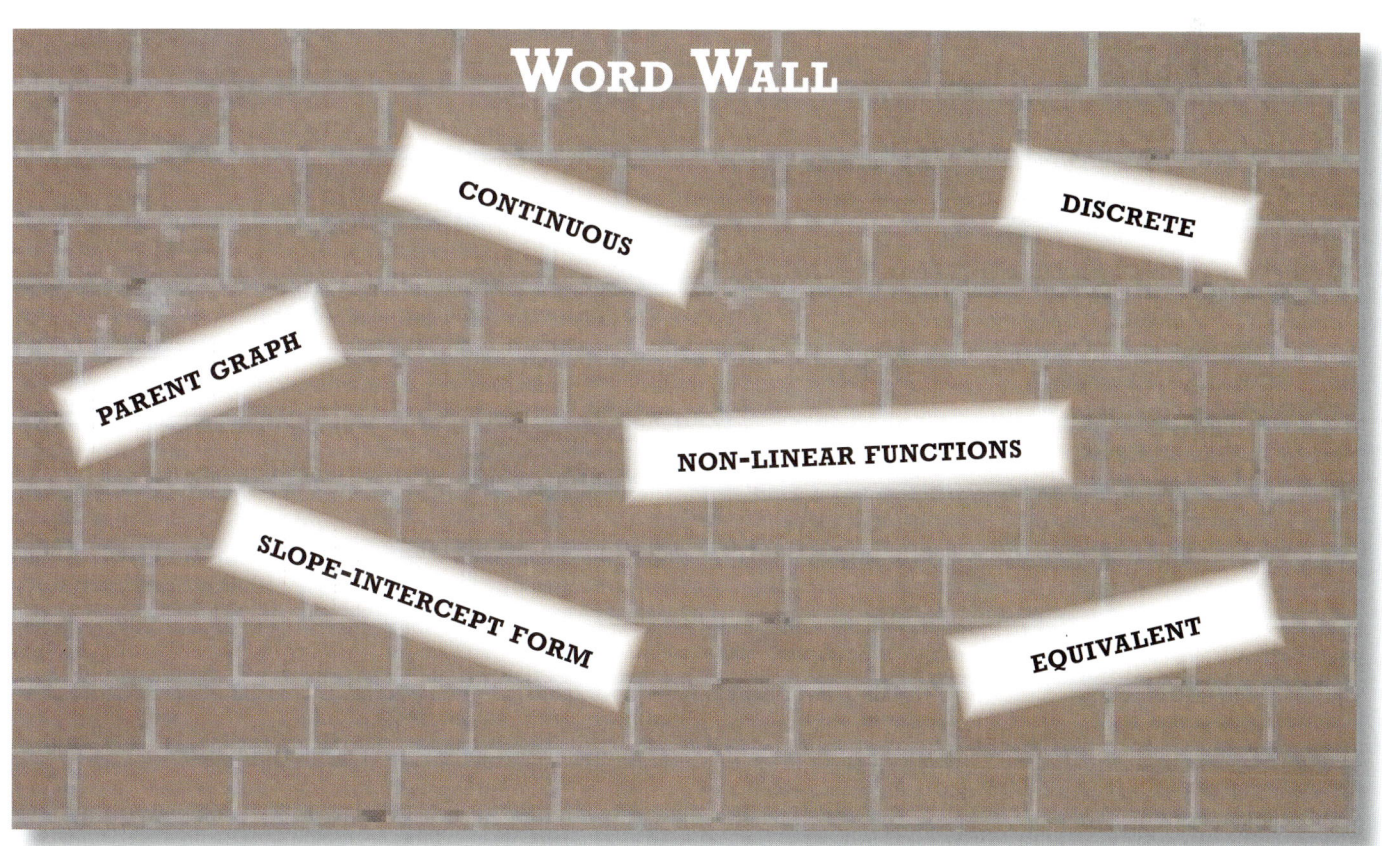

BLOCK 3 ~ USING LINEAR EQUATIONS
TIC-TAC-TOE

x- AND y-INTERCEPTS

Graph linear equations in standard form using the *x*- and *y*-intercepts.

See page 99 for details.

CLASS COMPETITION

Create a class competition where participants graph and write linear equations.

See page 114 for details.

GRAPHING DESIGN

Graph many linear equations to create an artistic design.

See page 133 for details.

CAREERS USING ALGEBRA

Research and write a report about different career choices where knowledge of algebra is essential.

See page 118 for details.

PARALLEL OR PERPENDICULAR LINES

Graph parallel and perpendicular lines. Make predictions about the slopes of the lines.

See page 105 for details.

QUADRATIC FUNCTIONS

Graph quadratic functions in factored form.

See page 129 for details.

EXPONENTIAL EQUATIONS

Use the exponential equation to predict future amounts.

See page 119 for details.

EQUIVALENT EQUATIONS

Make a flap book which matches pairs of equivalent linear equations.

See page 114 for details.

LINES OF BEST FIT

Draw lines of best fit, find equations and make predictions.

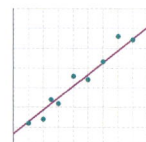

See page 110 for details.

GRAPHING USING SLOPE-INTERCEPT FORM

LESSON 3.1

 Graph linear equations in slope-intercept form.

In **Lesson 2.5** you were introduced to linear functions. Slope-intercept form is the most common equation used to represent a linear function. It is called this because the slope and the *y*-intercept are easily identified.

> **SLOPE-INTERCEPT FORM OF A LINEAR EQUATION**
> $$y = mx + b$$
> The slope of the line is represented by *m*.
> The *y*-intercept is *b*.

- Slope (*m*) is also called the rate of change. The slope gives you the rise over the run of the line.

- The *y*-intercept is also called the start value. The *y*-intercept is the location where the line crosses the *y*-axis. The ordered pair for the *y*-intercept will be (0, *b*).

- Equations in slope-intercept form may also be written $y = b + mx$.

Graphing an equation is a very important skill in mathematics because it is a visual representation of a mathematical equation. In this lesson you will learn how to graph an equation when it is presented in slope-intercept form.

EXAMPLE 1 Graph $y = \frac{1}{3}x - 2$. Clearly mark at least three points on the line.

SOLUTION First determine the slope and *y*-intercept. $y = \frac{1}{3}x - 2$ ← *y*-intercept
$m = \frac{1}{3}$ and $b = -2$ ← slope

Start by graphing the *y*-intercept on the coordinate plane.

Use the slope to find at least two more points. Remember rise over run.

Draw a straight line through the points. Put an arrow on each end.

EXAMPLE 2 Graph each linear equation. Clearly mark at least three points on each line.
  a. $y = 6 - 2x$ b. $y = -\frac{7}{2}x + 1$

SOLUTIONS

a. $y = 6 - 2x \rightarrow m = -2$ and $b = 6$

The slope is the coefficient of x.

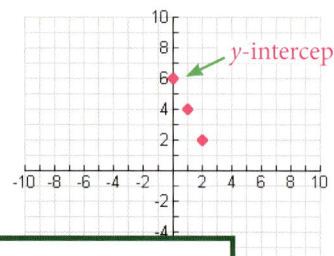

Graph the y-intercept of 6 and then graph 3 points using the slope fraction $-2 = \frac{-2}{1}$.

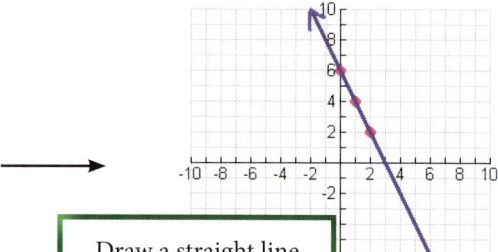

Draw a straight line through the points. Put an arrow on each end.

b. $y = -\frac{7}{2}x + 1 \rightarrow m = -\frac{7}{2}$ and $b = 1$

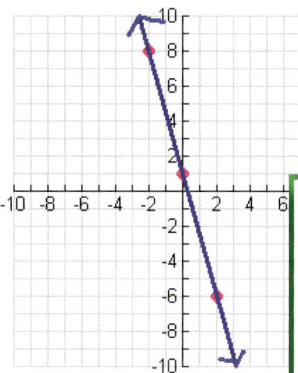

Sometimes you may have to use your slope in two directions. Go down 7 and right 2. In order to get another point on the coordinate plane, go up 7 and left 2.

$$\frac{-7}{2} = \frac{7}{-2}$$

As you learned in **Block 2**, there are lines that have a slope of zero and other lines that have an undefined slope. The equations of these lines are unique.

Zero Slope
y = constant

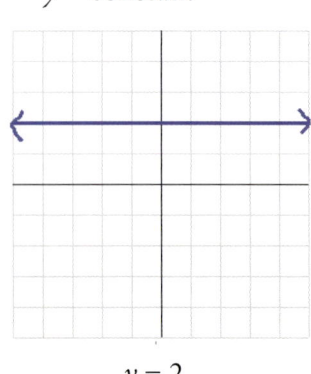

$y = 2$

Undefined Slope
x = constant

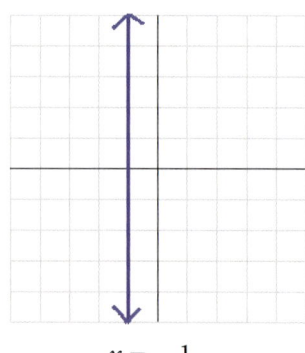

$x = -1$

When a graph can be drawn from beginning to end without lifting your pencil, it is **continuous**. Continuous graphs often represent real-world scenarios where something is happening continuously with no breaks. Some situations are modeled by linear equations but are not continuous. This could mean that it would not make sense to connect the points of the equation with a line.

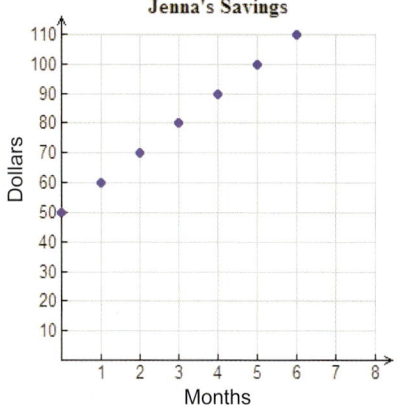

For example, Jenna started with $50 in her savings account. She adds $10 each month. The linear equation that represents the balance in her savings account is $y = 50 + 10x$. Since she only puts money in her account once a month, a line should not be drawn through the points on the graph. Only the whole numbers and 0 can be used as *x*-values. This graph is called **discrete** because it is represented by a unique set of points rather than a continuous line.

A graph can be continuous but limited to a certain quadrant or section of the graph. For example, Lamar types 40 words per minute. It would not make sense to graph points out of the first quadrant because he cannot type for a negative number of minutes or type a negative number of words. This graph is continuous because it can be drawn without lifting your pencil.

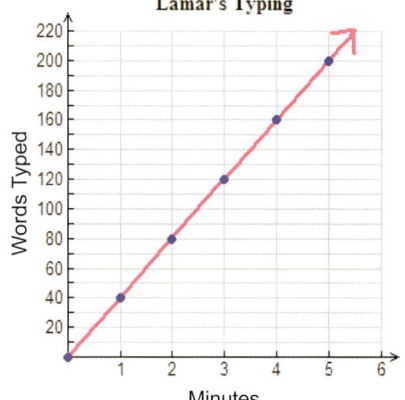

EXERCISES

Draw a coordinate plane for each problem. Graph the given equation. Clearly mark three points on the line.

1. $y = \frac{1}{2}x - 3$

2. $y = 1 - 3x$

3. $y = -\frac{2}{5}x + 6$

4. $y = x + 2$

5. $y = 4$

6. $y = -5 + \frac{4}{3}x$

7. $x = -2$

8. $y = 5x$

9. $y = -\frac{3}{2}x + 4$

10. Taylor's graphs of two different linear equations are seen below. His teacher told him both graphs were incorrect. Explain to Taylor why each of his graphs is not correct.

   a. $y = \frac{4}{5}x + 1$

   b. $x = 3$

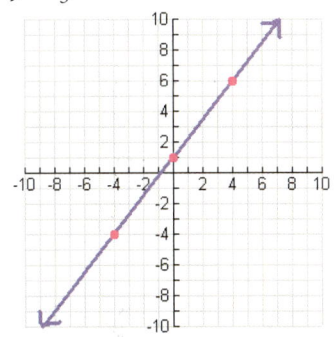
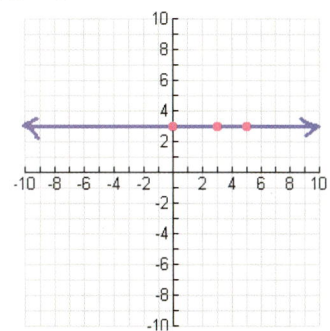

Lesson 3.1 ~ Graphing using Slope-Intercept Form

11. Daryl was given the linear equation $y = 1.5x + 2$. He was not sure how to graph this equation because its slope was a decimal. Follow the process Daryl decided to use.

x	y
−2	
0	
2	
4	

  a. Copy the table and use the equation to fill in the output values.
  b. Graph your ordered pairs (x, y) from the table on a coordinate plane. Draw a line through the points.
  c. Daryl was quite happy with the graph of the linear equation and was sure he had found the easiest way to deal with a linear equation which has a decimal slope value. Do you agree with him? Explain your reasoning.

12. Graph the three linear equations below on the same coordinate plane. Describe the similarities and differences of the three lines.

$$y = \tfrac{3}{4}x \qquad y = \tfrac{3}{4}x - 5 \qquad y = \tfrac{3}{4}x + 2$$

13. Create a linear equation that satisfies each condition. Graph your equations on a coordinate plane.
  a. Slope = $\tfrac{1}{3}$ and a negative y-intercept
  b. Slope = 0 and a y-intercept of 4
  c. A positive slope and a positive y-intercept
  d. A negative slope and a y-intercept of 0

14. Mrs. Samuels warned her class that the linear equations shown below were the most-often missed problems on the linear equations test she gave to her class last year. Explain why you think each problem might have been missed and then graph each equation.
  a. $y = x - 3$
  b. $y = 4$
  c. $x = -4$
  d. $y = 6 - x$

15. Falls City experienced a massive rainstorm from December 26th to December 30th in 1936. On the first day of the storm it rained 5.5 inches. It continued to rain 2.5 inches each day for the next four days.

x	y
1	
2	
3	
4	
5	

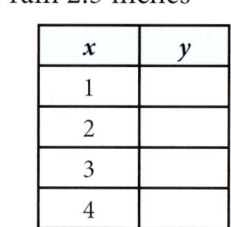

  a. Fill in the table with the TOTAL rain that had fallen during the storm as each day passed.
  b. In this situation, which number represents the slope (or rate of change)?
  c. Determine the y-intercept. Show all work necessary to justify your answer.
  d. Write a linear equation that represents the total rainfall in Falls City based on the number of days the storm had lasted.
  e. If the storm had continued at the same rate for 10 days, what would have been the total rainfall?

16. Dave was able to do 3 pull-ups before attending PE class. Each week of PE, the number of pull-ups he was able to do increased by 2.
  a. Write a linear equation representing the number of pull-ups, y, Dave was able to do in a given week, x.
  b. Would this graph be a continuous line? Why or why not?
  c. Graph the equation in the way that best models the situation.

Choose which of the following graphs is the best model for each situation. Explain your reasoning.

A. 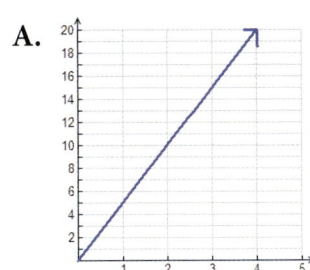   B. 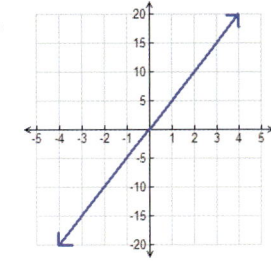   C.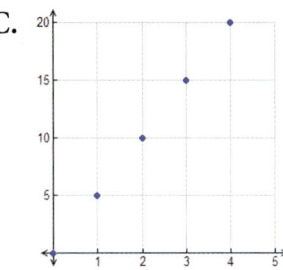

17. Ima started collecting coins. She adds 5 coins to her collection each week.

18. Todd runs at a rate of 5 miles per hour.

19. A continuous relationship where y is 5 times x.

REVIEW

Find the slope of the line that passes through the given points.

20. $(2, 9)$ and $(6, 11)$ **21.** $(1, 2)$ and $(1, 6)$ **22.** $(3, -1)$ and $(5, 0)$

23. $(-2, 6)$ and $(1, 9)$ **24.** $(4, 5)$ and $(0, 5)$ **25.** $(1, 2)$ and $(4, -3)$

Tic-Tac-Toe ~ x- and y-Intercepts

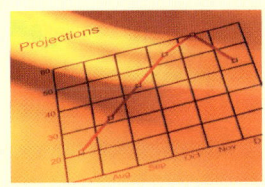

When linear equations are written in standard form ($Ax + By = C$), they can be graphed by finding the x- and y-intercepts and then drawing a line through those two points.
- In order to find the x-intercept, you must substitute 0 for y in the equation and then solve for x.
- In order to find the y-intercept, you must substitute 0 for x in the equation and then solve for y.

For example: Graph $2x - 3y = 6$ using the intercept method:

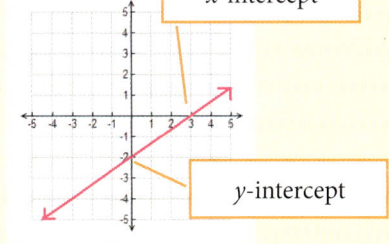

x-intercept	y-intercept
$2x - 3(0) = 6$	$2(0) - 3y = 6$
$2x = 6$	$-3y = 6$
$x = 3$	$y = -2$

Graph each of the following using the intercept method.

1. $5x + 4y = 20$ **2.** $-3x + y = 6$ **3.** $2x - 3y = 9$
4. $-3x + 5y = 15$ **5.** $4x - y = -8$ **6.** $4x + 7y = 14$
7. Convert each of the equations in **#1-6** into slope-intercept form using the method shown in **Lesson 3.4**. Use the slope-intercept equation to verify that each graph has the correct y-intercept and slope.

WRITING LINEAR EQUATIONS FOR GRAPHS

LESSON 3.2

 Write a linear equation for a given graph.

EXPLORE! **FIND THE EQUATION**

Stacey and Mario like to go to the coffee shop before school. They decided to conduct an experiment to study the rate at which their coffees cool when left untouched on the table. The graph below shows the information they gathered.

Step 1: What is the real world meaning of the point (0, 160)? How about the point (10, 120)?

Step 2: Use the slope formula, $\frac{y_2 - y_1}{x_2 - x_1}$, to find the slope of the line. Does it matter which points from the graph you use in the formula?

Step 3: What is the real-world meaning of the slope?

Step 4: What is the y-intercept of this graph?

Step 5: Write an equation in slope-intercept form that represents this graph.

Step 6: Use your equation to determine the temperature of the coffee after 12 minutes.

Step 7: According to Stacey and Mario's experiment, the coffee continued to cool at the same rate every minute that passed. Do you think the coffee will continue to cool at this rate if the coffee is left on the table for one hour? Verify your theory using your equation.

A linear equation can be written for a specific line if you know the slope and y-intercept. The y-intercept can be determined by locating the point where the graph crosses the y-axis $(0, b)$. The slope must be calculated using a slope triangle or the slope formula. Remember that when dealing with real-world graphs, the y-intercept is referred to as the start value and the slope is called the rate of change.

WRITING A LINEAR EQUATION FROM A GRAPH

1. Locate the y-intercept on the graph.
2. Find the slope of the line.
3. Write the equation in slope-intercept form, $y = mx + b$.

EXAMPLE 1 Determine the slope and *y*-intercept of each graph. Write the equation for each graph in slope-intercept form.

a. 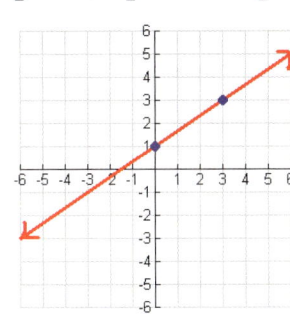 b. 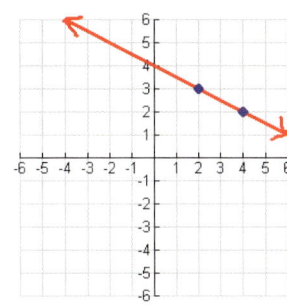 c.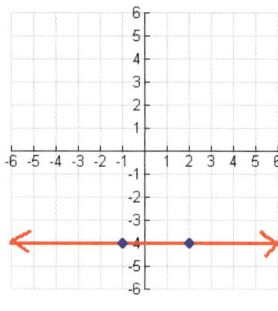

SOLUTIONS

a. The line crosses the *y*-axis at 1 so $b = 1$.
The slope triangle shows that $m = \frac{2}{3}$.
The equation in slope-intercept form is: $y = \frac{2}{3}x + 1$.

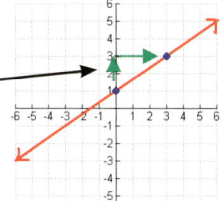

b. The line crosses the *y*-axis at 4 so $b = 4$.
The slope triangle shows that $m = -\frac{1}{2}$.
The equation in slope-intercept form is: $y = -\frac{1}{2}x + 4$.

c. The line crosses the *y*-axis at −4 so $b = -4$.
The slope triangle shows that $m = \frac{0}{3} = 0$.
The equation in slope-intercept form is $y = 0x - 4$ which can also be written as $y = -4$.

It is very useful to have equations for graphs that represent real-world situations because you can use the equation to predict future or past data.

EXAMPLE 2 Zach enjoys running each day after school. The graph below represents the distance Zach has traveled based on the number of minutes he has been running.

a. Find the slope-intercept equation that represents the situation shown on the graph.
b. Use your equation to determine how far Zach will have gone in 28 minutes.
c. Use your equation to determine how long it will take Zach to run 10 miles.

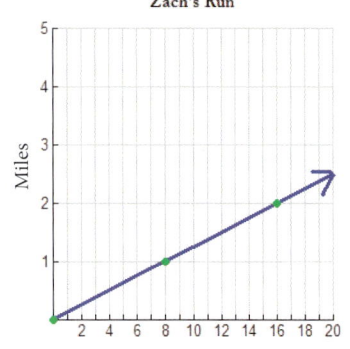 

SOLUTIONS

a. The line crosses the *y*-axis at 0 so $b = 0$.
The two marked points are (8, 1) and (16, 2). Use the slope formula to calculate the slope.

$$\frac{y_2 - y_1}{x_2 - x_1} = \frac{2 - 1}{16 - 8} = \frac{1}{8}$$

The slope-intercept equation is $y = \frac{1}{8}x + 0$ or $y = \frac{1}{8}x$.

Lesson 3.2 ~ Writing Linear Equations for Graphs

EXAMPLE 2
SOLUTIONS
(CONTINUED)

b. Since the x-values represent minutes, substitute 28 for x to determine how far Zach runs in 28 minutes:

$$y = \tfrac{1}{8}x$$
$$y = \tfrac{1}{8}(28)$$
$$y = \tfrac{28}{8} = 3\tfrac{4}{8} = 3\tfrac{1}{2} = 3.5$$

Zach runs 3.5 miles in 28 minutes.

c. Since the y-values represent miles, substitute 10 for y to determine how long it will take Zach to run 10 miles.

$$10 = \tfrac{1}{8}x$$
$$\tfrac{8}{1} \cdot 10 = \tfrac{1}{8}x \cdot \tfrac{8}{1}$$
$$80 = x$$

Multiply by the reciprocal.

Zach's 10 mile run will take 80 minutes (one hour and twenty minutes).

Stacey and Mario found it was easy to write a slope-intercept equation from the graph because all they had to do was find the y-intercept and slope and put it into the form $y = mx + b$. They found the equation for their coffee experiment in the *Explore!* to be $y = -4x + 160$. They tested their formula by substituting values for the temperature of the coffee to see if it matched the number of minutes shown on the graph.

Temperature = 150° F → Since the y-values represent the temperature, substitute 150 for y.

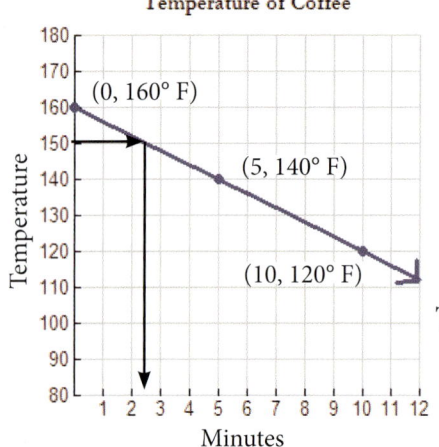

$$150 = -4x + 160$$
$$-160 \qquad -160$$
$$\tfrac{-10}{-4} = \tfrac{-4x}{-4}$$
$$2.5 = x$$

The coffee was 150° after 2.5 minutes.

Stacey and Mario used their formula to predict when their coffee would freeze.
Temperature = 32° F (freezing) → Since the y-values represent temperature, substitute 32 for y.

$$32 = -4x + 160$$
$$-160 \qquad -160$$
$$\tfrac{-128}{-4} = \tfrac{-4x}{-4}$$
$$32 = x$$

According to the equation, the coffee would freeze in 32 minutes.

Stacey and Mario decided their formula only works for the first 20 minutes or so. It is not likely that coffee is going to reach a freezing temperature while it is sitting on the table.

Lesson 3.2 ~ Writing Linear Equations for Graphs

EXERCISES

Identify the slope and *y*-intercept of each graph. Write the corresponding linear equation in slope-intercept form.

1.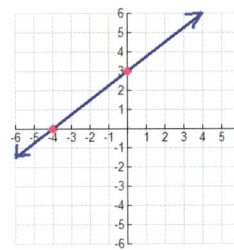
2.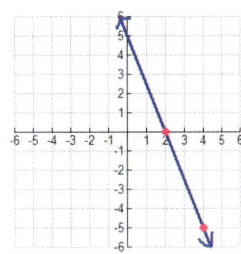
3.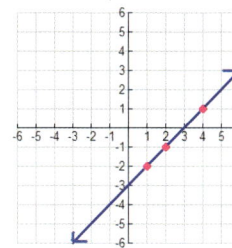
4.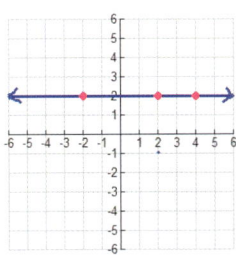
5.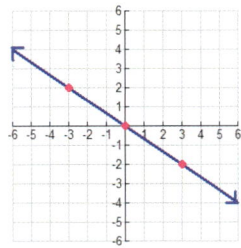
6.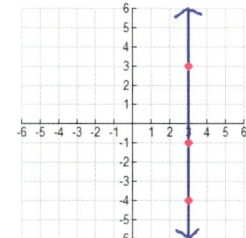

7. Skyler's total savings are shown on the graph.
   a. Find the slope-intercept equation that represents the graph.
   b. Determine how much Skyler will have in his savings account after 18 months.
   c. Determine how many months it will take before Skyler has $212 in his savings. Show all work necessary to justify your answer.

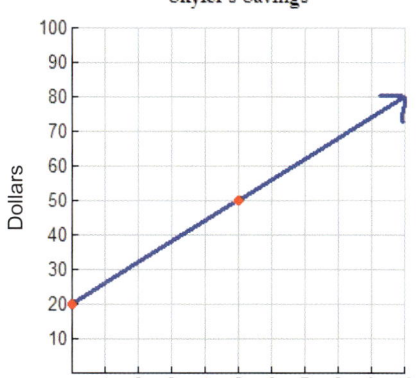

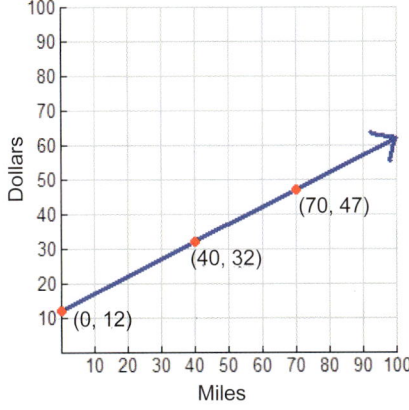

8. Javier owns a car rental company. He provides the graph seen at left for his customers to see the price for renting a sedan based on the number of miles they drive.
   a. Find the equation (in slope-intercept form) that represents the amount Javier charges based on the number of miles driven.
   b. Determine the amount a customer will have to pay if she rents a sedan and drives it 120 miles.
   c. Leticia rented a sedan from Javier. When she returned it, her bill was $38.50. How many miles did she drive? Show all work necessary to justify your answer.

9. What is the minimum number of points needed on a linear graph to find the equation for the line? Explain your reasoning.

10. What type of linear equation does not have a *y*-intercept? Give an example of a graph and its corresponding equation that fit this category of linear equations.

11. At two different times during the summer, Kirsten measured the height of a sunflower she had planted in May. She measured it when she first planted the flower and then again 3 weeks later.

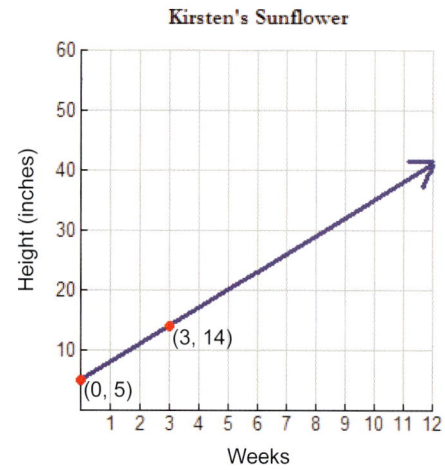

    a. Find the slope-intercept equation that represents the height of Kirsten's flower based on the number of weeks since she planted it.
    b. Determine exactly how tall the sunflower will be after 8 weeks.
    c. Determine how many weeks have passed if the plant is 47 inches tall.

12. Luke drained his 20-gallon fish tank. At two different times, he measured the amount of water left in the tank. He graphed the information on the graph shown at right.

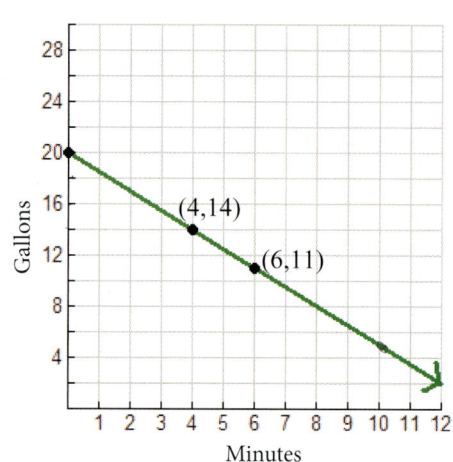

    a. Find the slope-intercept equation that represents the number of gallons left in the fish tank since he began draining it.
    b. Use your equation to determine how much water will be left in the tank after 10 minutes.
    c. When will the water be completely drained from the tank? Show all work necessary to justify your answer.

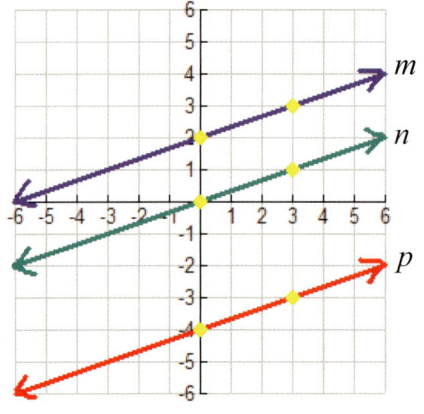

13. Use the graph at left to answer the following questions.
    a. Find the slope-intercept equations for lines *m*, *n* and *p*.
    b. What do the three equations have in common?
    c. What geometry term can be used to describe the relationship between these three lines?

REVIEW

14. Evaluate the following expressions when $x = 3$ and $y = -4$.
   a. $2x + 5y$
   b. $-4x - 6y$
   c. $\frac{1}{2}y - 5x$

State whether each equation is true or false for the values of the variables given.

15. $y = 3x + 1$ when $x = 2$ and $y = 6$

16. $y = \frac{4}{3}x - 4$ when $x = 6$ and $y = 4$

17. $y = -2x + 7$ when $x = 5$ and $y = -3$

18. $y = \frac{1}{4}x$ when $x = 10$ and $y = 2$

19. $y = 4 - x$ when $x = 3$ and $y = 1$

20. $y = 2 - \frac{1}{2}x$ when $x = 1$ and $y = 2$

21. Ali's resting metabolic rate is 1,320 calories per day. She goes for a 6-mile run in the morning and burns calories at a rate of 100 calories per mile. After school she goes bowling. If she burns 3 calories per minute, approximately how many minutes will she need to bowl to burn a total of 2,000 calories for the day? Show all work necessary to justify your answer.

Tic-Tac-Toe ~ Parallel or Perpendicular Lines

Perpendicular lines are lines that intersect at a 90° angle. Parallel lines never intersect. Each pair of lines given below is either parallel or perpendicular.

SET #1	SET #2	SET #3	SET #4	SET #5
$y = 2x + 3$	$-3x + 2y = 6$	$y - x = 5$	$-x + 2y = -4$	$y = \frac{1}{3}x + 3$
$y = 2x - 4$	$y = -\frac{2}{3}x - 4$	$y = x - 2$	$4x + 2y = 8$	$y = -3x - 1$

Step 1: If necessary, convert each equation into slope-intercept form using the method shown in **Lesson 3.4**.

Step 2: Graph each pair of equations on the same coordinate plane.

Step 3: State whether each pair of lines is parallel or perpendicular.

Step 4: After completing all five sets of graphs, develop a hypothesis on how to use the slope-intercept equation to determine if lines are parallel or perpendicular without graphing. Explain how you arrived at your hypothesis.

WRITING LINEAR EQUATIONS FROM KEY INFORMATION

LESSON 3.3

 Write a linear equation in slope-intercept form when given information about the line.

There are two pieces of information you need to be able to write an equation in slope-intercept form: the slope and the *y*-intercept. You learned how to determine the equation when given a graph of the linear equation. In this lesson you will find equations for specific lines when given different pieces of information about the lines.

When you are given the slope and the *y*-intercept for a line you need to insert the information into $y = mx + b$ for the appropriate variables.

EXAMPLE 1   Write the equation of a line that has a slope of −2 and *y*-intercept of 5.

SOLUTION

Write the general slope-intercept equation. $y = mx + b$

Substitute −2 for *m* since *m* represents the slope. $y = -2x + b$

Substitute 5 for *b* since *b* represents the *y*-intercept. $y = -2x + 5$

The equation is $y = -2x + 5$.

When you are not directly given the slope and *y*-intercept, there are steps you can follow to find both the slope and *y*-intercept. Once you have both the slope and *y*-intercept, you can write a linear equation in slope-intercept form.

WRITING A LINEAR EQUATION WHEN GIVEN KEY INFORMATION

1. Find the slope (*m*) of the line.
2. Find the *y*-intercept (*b*) of the line. If necessary, substitute the slope for *m* and one ordered pair (*x*, *y*) for the corresponding variables in the equation $y = mx + b$. Solve for *b*.
3. Write the equation in the form $y = mx + b$.

EXAMPLE 2

Write the equation of a line that has a slope of $\frac{4}{3}$ and goes through the point (−3, 1).

SOLUTION

The slope is given.

$$m = \frac{4}{3}$$

Write the slope-intercept equation with the slope.

$$y = \frac{4}{3}x + b$$

Find the *y*-intercept, *b*, by substituting the given point (−3, 1) for *x* and *y* in the slope-intercept equation.

$$1 = \frac{4}{3}(-3) + b$$
$$1 = -4 + b$$
$$+4 \quad +4$$

Solve for *b*.

$$5 = b$$

Write the equation by substituting *m* and *b*.

$$y = \frac{4}{3}x + 5$$

Check by graphing.

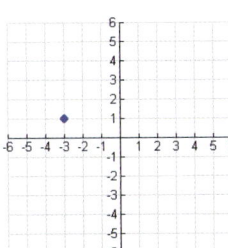
Plot the point (−3, 1).

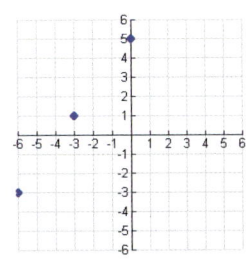
Use the slope $\frac{4}{3}$ to find at least two more points.

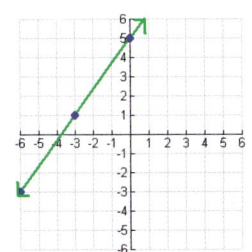
Draw a line through the points.

If you are given two points on a line, first find the slope using the slope formula or a slope triangle. Then follow the process in **Example 2** to write the slope-intercept equation.

EXPLORE!

TRIANGLE LINES

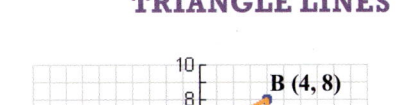

A triangle consists of three line segments. A segment is a portion of a line.

Step 1: Find the equation of the line that contains $\overline{AB}$.

Step 2: Find the equation of the line that contains $\overline{AC}$.

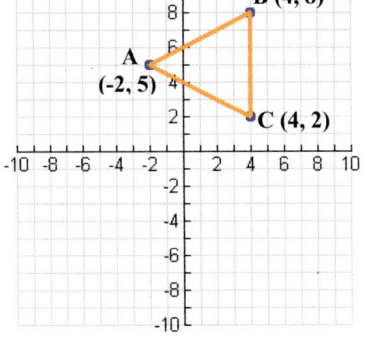

Step 3: Find the equation of the line that contains $\overline{BC}$.

Step 4: Ryan made his own triangle. He chose three points and wants you to find the equations of the three lines that make up his triangle.

His points are (3, 3), (−2, −7) and (−7, −2). Can you find the three linear equations that intersect to make his triangle?

Step 5: The points where the line segments meet in a triangle are called the vertices. Graph your three lines on a piece of graph paper using the slope and *y*-intercepts from your equations. Does the triangle that is formed have the same three vertices that Ryan chose?

Lesson 3.3 ~ Writing Linear Equations from Key Information

EXERCISES

Write an equation in slope-intercept form when given the slope and y-intercept.

1. slope = $\frac{6}{5}$, y-intercept = 8

2. slope = −4, y-intercept = 1

3. slope = 1, y-intercept = 2

4. slope = 0, y-intercept = −3

5. In 2010, the population of South Carolina was approximately 4,600,000. During the next two years, the population increased by approximately 45,000 people each year.
   a. Write an equation in slope-intercept form that represents the population, y, of South Carolina in terms of the number of years, x, since 2010.
   b. Estimate the population of South Carolina in 2020 if this trend continues. Show all work necessary to justify your answer.

Write an equation in slope-intercept form when given the slope and one point on the line.

6. slope = 2, goes through the point (1, −4)

7. slope = $-\frac{3}{4}$, goes through the point (4, 2)

8. slope = −1, goes through the point (−2, 3)

9. slope = $\frac{5}{2}$, goes through the point (−6, −10)

10. slope = $\frac{1}{2}$, goes through the point (3, 4)

11. slope = 0, goes through the point (11, 8)

12. One New York City taxi company charges an initial fee plus $0.50 for each minute of the ride. Tammy was in the taxi for 15 minutes. The cost was $10.50. Let x represent the number of minutes and y represent the total cost of the taxi ride.
    a. Identify the slope and one ordered pair from the information given.
    b. Find the equation of the line that fits this information. Show all work necessary to justify your answer.
    c. Joe uses this taxi company for a 30 minute ride. How much should he expect to pay?

Write an equation in slope-intercept form when given two points.

13. goes through the points (1, 2) and (3, 8)

14. goes through the points (−5, 9) and (4, 0)

15. goes through the points (6, 9) and (−3, 6)

16. goes through the points (8, −4) and (−2, 1)

17. goes through the points (10, 6) and (0, −2)

18. goes through the points (4, −5) and (4, −1)

19. Evan says it is impossible to create an equation for two points that have the same x-values, like (1, 5) and (1, 8), because the slope is undefined. Do you agree or disagree? Explain your reasoning.

20. Find the equation of the line that goes through the points (−2, 7) and (−1, 5). Support your answer by substituting both ordered pairs into your equation to show they make the equation true.

Lesson 3.3 ~ Writing Linear Equations from Key Information

21. At 2 weeks old, Bob's baby sister weighed 9 pounds. When she was 8 weeks old, she weighed 12 pounds. Let x represent how old the baby is in weeks and y represent the baby's weight in pounds.
  a. Write two ordered pairs that use the data about Bob's sister.
  b. Find the equation of the line that goes through these two points.
  c. If Bob's sister continues to grow at this rate, how much will she weigh when she is 20 weeks old? Is this reasonable? Why or why not?

22. Four line segments make the four sides of a quadrilateral on the coordinate plane below. Find the equations of the lines containing each side: $\overline{AB}$, $\overline{BC}$, $\overline{CD}$, $\overline{AD}$. Are there any similarities in the equations for lines $\overline{AB}$ and $\overline{CD}$? How about $\overline{BC}$ and $\overline{AD}$?

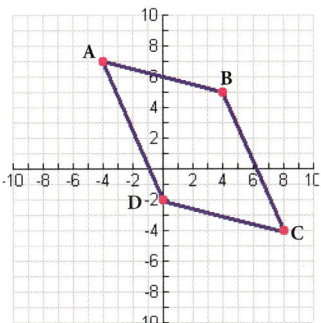

23. A raft rental company on the Gauley River rents rafts for a set fee plus an additional charge per hour. Francis asked two different people how many hours they had rented their rafts for and how much it cost. One rented a raft for 6 hours and paid $32. Another rented a raft for 11 hours and paid $47. Let x represent the length of time in hours a raft is rented for and let y represent the total cost.
  a. What is the linear equation that represents the data Francis collected?
  b. What number in the linear equation represents the amount of the set fee?
  c. What is the real-world meaning of the slope in this equation?
  d. How much will someone pay for a raft rental from this company if he only keeps the raft for 4 hours? Show all work necessary to justify your answer.

REVIEW

Write the linear equation for each graph in slope-intercept form.

24.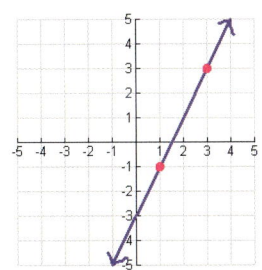

25.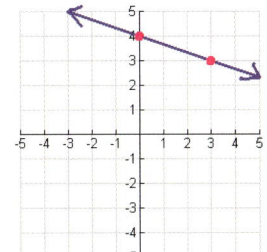

26.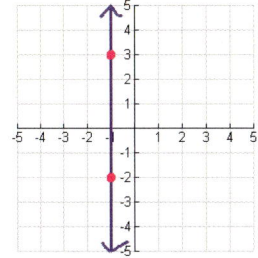

Draw a coordinate plane for each problem. Graph each equation. Clearly mark three points on the line.

27. $y = \frac{5}{3}x - 4$

28. $y = -4x + 5$

29. $x = 3$

Write the slope-intercept equation that matches the data.

30.

Hours	Distance Traveled
0	45
1	75
2	105
3	135
4	165

31.

Days	Plant Height
1	3.3
2	3.6
6	4.8
8	5.4
10	6.0

32.

Cups of Coffee Sold	Profit $
0	−15
4	−7
9	3
15	15
100	185

TIC-TAC-TOE ~ LINES OF BEST FIT

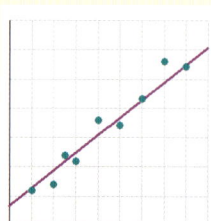

You have learned how to take real-world data and graph the data points on a scatter plot. When real-world data is collected there is usually not a single line that passes through all of the data points, but you can often see a linear pattern on the scatter plot. In this activity, you will find a line that best fits the data you are given. This line is called the line of best fit. The steps to finding a line of best fit are shown in the box below.

FINDING A LINE OF BEST FIT

1. Plot the data points and determine the general direction of the data.
2. Draw a line in that general direction that has approximately the same number of points above and below the line.
3. Locate two points on your line. Approximate the x- and y-coordinates for each point (these do not need to be data points from the original data).
4. Find the slope-intercept equation of the line.

The data in the table shows the fare costs for different lengths of rides on the city bus.

1. Graph the data points.

2. Draw a line of best fit.

3. Find the equation for the line of best fit.

4. Use your equation to predict the cost to ride this city bus 50 miles.

5. Use your equation to predict how far you could ride the bus for $10.

Miles	Cost
2	$1.00
5	$1.75
10	$2.00
12	$2.50
16	$2.50
20	$3.25
26	$3.50
30	$4.75

DIFFERENT FORMS OF LINEAR EQUATIONS

LESSON 3.4

 Convert different forms of linear equations to slope-intercept form.

Linear equations come in many different forms. So far in this book, you have used the slope-intercept form. The slope-intercept form gives both the slope and y-intercept of the graph. In this lesson, you will work with linear equations in other forms and convert them into slope-intercept form in order to graph the equations.

STANDARD FORM: $Ax + By = C$ where A and B are not both zero.
POINT-SLOPE FORM: $y - y_1 = m(x - x_1)$ or $y = m(x - x_1) + y_1$

It is important to know how to change equations into forms you can recognize and use. The key to converting a linear equation into slope-intercept form is to remove all parentheses using the Distributive Property and then isolating the y-variable.

Converting STANDARD FORM to SLOPE-INTERCEPT FORM

Process	Example: $-3x + 4y = 12$
Step 1: Move the term containing the x-value to the other side of the = sign with the constant.	$-3x + 4y = 12$ $+3x \qquad\qquad +3x$
Step 2: Re-write the equation. Remember that you cannot combine terms unless they are like terms.	$4y = 12 + 3x$
Step 3: Isolate y by dividing by the coefficient on the y-value. Divide EVERY TERM by the coefficient of y.	$\frac{4y}{4} = \frac{12 + 3x}{4}$ $y = 3 + \frac{3}{4}x$
Step 4: Re-write the equation. Leave the coefficient of x as a simplified fraction when you divide because this is the slope of the graph.	or $y = \frac{3}{4}x + 3$

EXAMPLE 1 Convert the following from STANDARD FORM to SLOPE-INTERCEPT FORM.
a. $-5x + 2y = -20$ 	b. $x - 3y = 9$

Remember that $-x = -1x$.

SOLUTIONS

a. $-5x + 2y = -20$
$+5x \qquad\qquad +5x$
$\frac{2y}{2} = \frac{-20 + 5x}{2}$
$y = -10 + \frac{5}{2}x$
or
$y = \frac{5}{2}x - 10$

b. $x - 3y = 9$
$-x \qquad\qquad -x$
$\frac{-3y}{-3} = \frac{9 - x}{-3}$
$y = -3 + \frac{1}{3}x$
or
$y = \frac{1}{3}x - 3$

Converting POINT-SLOPE FORM to SLOPE-INTERCEPT FORM

Process	Example: $y - 5 = 2(x + 1)$
Step 1: Use the Distributive Property to remove any parentheses.	$y - 5 = 2(x + 1) \rightarrow y - 5 = 2x + 2$
Step 2: Combine like terms that are on the same side of the equals sign.	Not necessary in this example.
Step 3: Get y "by itself" by balancing the equation using the Properties of Equality.	$\begin{array}{r} y - 5 = 2x + 2 \\ +5 +5 \end{array}$
Step 4: Write final answer in the form $y = mx + b$.	$y = 2x + 7$

EXAMPLE 2 Convert the following from POINT-SLOPE form to SLOPE-INTERCEPT form.
  a. $y = \frac{1}{2}(x - 4) + 1$ b. $y + 4 = -3(x - 2)$

SOLUTIONS

a. $y = \frac{1}{2}(x - 4) + 1$
$y = \frac{1}{2}x - 2 + 1$ *Combine like terms.*
$y = \frac{1}{2}x - 1$

b. $y + 4 = -3(x - 2)$ *Balance the equation.*
$y + 4 = -3x + 6$
$ -4 -4$
$y = -3x + 2$

In **Example 2**, the equation $y = \frac{1}{2}(x - 4) + 1$ is **equivalent** to $y = \frac{1}{2}x - 1$. That means that both equations represent the same line even though the equations look different. Linear equations can come in many forms, but every linear equation can be converted to an equivalent slope-intercept equation. Try the Explore! to practice converting all types of equations to slope-intercept form.

CONVERTING EQUATIONS TO SLOPE-INTERCEPT FORM

1. Use the Distributive Property to remove any parentheses.
2. Combine all like terms on the same side of the equals sign.
3. Isolate y by balancing the equation using the Properties of Equality.
4. Re-write the equation in slope-intercept form.

EXPLORE!   **ONE OF THESE THINGS**

In each set of four equations "one of these things is not like the other". Three of the linear equations in each set are equivalent and one is not. For each set, find the three that are similar and give the slope-intercept form that they are equivalent to. Graph that equation on a coordinate plane.

SET 1
$y + 10 = 3(x + 2)$
$y = 1 + 3(x + 1)$
$-9x + 3y = -12$
$6x - 2y = 8$

SET 2
$2x + 4y = 4$
$y = \frac{1}{2}(x + 6) + 4$
$y + 1 = -\frac{1}{2}(x - 4)$
$5x + 10y = 10$

EXERCISES

Match each equation to its equivalent equation in slope-intercept form.

1. $y + 6 = 3(x + 2)$
2. $y = \frac{1}{2}(x + 8) - 2$
3. $y + 1 = 1(x - 3)$
4. $-4x + y = -2$
5. $2x - 4y = -4$
6. $2x + 4y = 8$

A. $y = 4x - 2$
B. $y = \frac{1}{2}x + 2$
C. $y = 3x$
D. $y = -\frac{1}{2}x + 2$
E. $y = x - 4$
F. $y = \frac{1}{2}x + 1$

Convert each equation to slope-intercept form.

7. $y + 3 = 4(x + 6)$
8. $6x + 2y = 12$
9. $y = -2 + \frac{1}{3}(x + 9)$
10. $2x - 5y = -15$
11. $-x - 2y = 2$
12. $y - 1 = -2(x - 5)$
13. $y = \frac{3}{4}(x + 12) - 2$
14. $-7x + y = 6$
15. $y + 15 = 4(x + 6)$

One of the two equations listed in each problem matches the graph. Determine which equation is represented by the graph. Explain how you know your answer is correct.

16. $y - 1 = 2(x + 1)$
    OR
    $6x + 3y = 9$

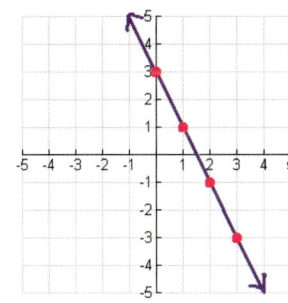

17. $6x + 2y = -4$
    OR
    $y = \frac{1}{3}(x - 9) + 1$

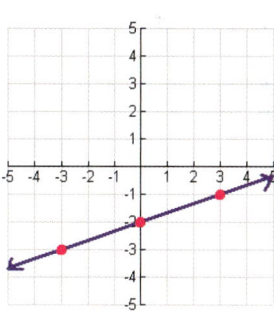

REVIEW

Write an equation in slope-intercept form that satisfies the information given about the line.

18. has a slope of $\frac{5}{2}$ and a y-intercept of 3
19. has a slope of -3 and goes through the point $(3, 1)$
20. has a slope of 5 and goes through the origin
21. goes through the points $(6, 1)$ and $(10, -1)$
22. goes through the points $(-2, 5)$ and $(4, 11)$
23. has a slope of 0 and a y-intercept of -5
24. goes through the points $(4, 1)$ and $(4, 9)$
25. goes through the points $(1, -3)$ and $(2, -6)$

Tic-Tac-Toe ~ Equivalent Equations

Step 1: On a sheet of notebook paper, write 10 linear equations in point-slope or standard form. Next to each equation, write the equivalent linear equation in slope-intercept form.

Step 2: Take a blank sheet of white paper and fold it down the middle vertically. Cut the front sheet into equal-sized sections.

Step 3: Write a linear equation from your list that is not in slope-intercept form on each flap. Inside the flap, show the steps to reach the equivalent equation that is in slope-intercept form.

Tic-Tac-Toe ~ Class Competition

Create a game that can be played with the whole class as a review for the material in this Block. Create a "Teacher's Guide" with the four categories shown below. In each category, generate three questions that vary in level of difficulty from easiest to hardest. Include the answer for each question in the "Teacher's Guide".

Writing Equations from Graphs	Writing Equations from Key Information	Graphing Linear Equations	Equivalent Equations
Level 1	Level 1	Level 1	Level 1
Level 2	Level 2	Level 2	Level 2
Level 3	Level 3	Level 3	Level 3

Create a document explaining the class competition rules. Here are some things to consider:

- How many teams should there be?
- How are the teams decided?
- How do the teams answer the questions (as individuals or as a team)?
- How much time does a team have to answer a question?
- What happens if a team gets the answer wrong?
- How do teams score points?
- When is the game over?

MORE GRAPHING LINEAR EQUATIONS

LESSON 3.5

 Graph linear equations that are not written in slope-intercept form.

EXPLORE! **MATCH ME**

Jared and Wendy are playing a matching game. Each card in the deck has either a graph or a linear equation on it. The goal of the game is to be the first to match the six Equation Cards to their corresponding Graph Cards.

Step 1: Convert each Equation Card to slope-intercept form.

Step 2: Match each Equation Card to its corresponding Graph Card.

1. $y + 6 = -2(x - 5)$

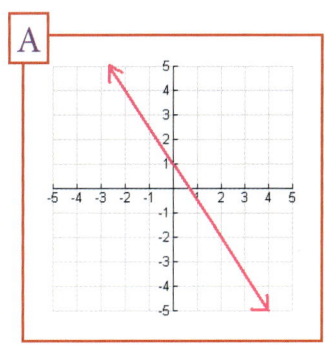
A.

2. $6y = -12$

B.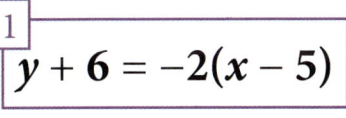

C.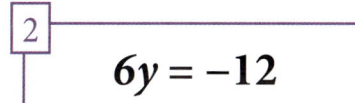

3. $y = -\frac{1}{3}(x + 3) + 1$

4. $y - 5 = 2(x - 4)$

D.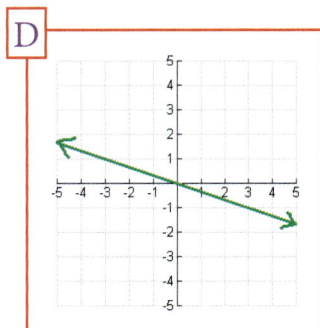

5. $3x + 2y = 2$

E.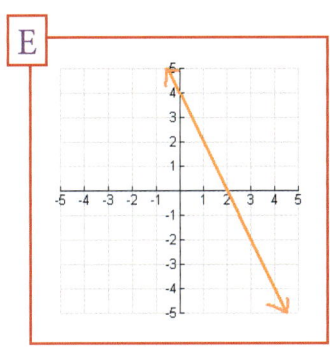

F.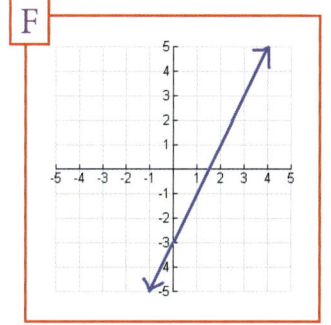

6. $-3x + 3y = -12$

Step 3: On your own paper, create two more Equation Cards and their corresponding graph cards that could be used in a future matching game.

Lesson 3.5 ~ More Graphing Linear Equations

No matter what form a linear equation is written in, it can be converted to slope-intercept form and graphed. Remember that the key steps to converting an equation into slope-intercept form include using the Distributive Property and isolating the *y*-variable.

EXAMPLE 1 Convert the following equations to slope-intercept form and then graph.
a. $4x + 3y = 12$
b. $y = \frac{1}{2}(x - 4) + 3$

SOLUTIONS

a. $\cancel{4x} + 3y = 12$
$-4x -4x$
$\cancel{3}y = 12 - 4x$
$\cancel{3} 3$
$y = 4 - \frac{4}{3}x$

b. $y = \frac{1}{2}(x - 4) + 3$
$y = \frac{1}{2}x - 2 + 3$
$y = \frac{1}{2}x + 1$

Every line has an infinite number of points that make up the line. In certain situations, verification is needed to determine whether or not a point lies on a certain line. In order to determine this, the equation DOES NOT need to be graphed. The *x*- and *y*-values from the ordered pair can be substituted for the *x*-and *y*-variables in the linear equation. If the values make the equation true, then the point lies on the line.

EXAMPLE 2 Determine if each point is on the given line.
a. Is the point $(3, -4)$ on the line $5x + 2y = 7$?
b. Is the point $(0, 5)$ on the line $y + 3 = 2(x + 3)$?

SOLUTIONS

a. Substitute 3 for *x* and -4 for *y*.

$5x + 2y \stackrel{?}{=} 7$
$5(3) + 2(-4) \stackrel{?}{=} 7$
$15 + -8 \stackrel{?}{=} 7$
$7 = 7$

The point is on this line.

b. Substitute 0 for *x* and 5 for *y*.

$y + 3 \stackrel{?}{=} 2(x + 3)$
$5 + 3 \stackrel{?}{=} 2(0 + 3)$
$8 \stackrel{?}{=} 2(3)$
$8 \neq 6$

The point is NOT on this line.

EXERCISES

Convert each equation to slope-intercept form and graph. Clearly mark at least three points on each line.

1. $-5x + 2y = -8$
2. $y - 8 = -3(x + 1)$
3. $y = 1 + \frac{1}{2}(x + 4)$
4. $x + 4y = 12$
5. $4x - 3y = 6$
6. $y = 2(x + 5) - 8$
7. $4x = 8$
8. $y = \frac{3}{4}(x + 8) - 9$
9. $y + 6 = 3(x + 2)$
10. $y - 11 = -\frac{5}{2}(x + 2)$
11. $-x + y = -6$
12. $-5y = 20$

13. Write a linear equation that is not in slope-intercept form. Convert it to slope-intercept form and graph it. Show that your linear equations are equivalent by substituting one point from your graph into both equations to show that they make each equation true.

14. Patti signed up for a cell phone plan that charges an initial monthly fee and a set rate per minute she talks on the phone each month. The equation she was given to calculate her total bill, y, was $y = 0.1(x + 20) + 7$ where x represents the number of minutes she talks on the phone in one month.
    a. Convert the equation into slope-intercept form.
    b. What is Patti's initial fee each month?
    c. What is her rate per minute?
    d. Last month, Patti talked on the phone for 427 minutes. How much was her total bill for last month? Show all work necessary to justify your answer.

15. Is the point $(-1, 4)$ on the line $3x + 2y = 5$?

16. Is the point $(6, 0)$ on the line $y = -11 + 2(x - 1)$?

17. Is the point $(0, 0)$ on the line $y = -\frac{1}{2}(x + 4) + 2$?

18. Is the point $(2, 10)$ on the line $5x - y = 0$?

19. Is the point $(-6, -2)$ on the line $-x + 2y = -10$?

20. Is the point $(-\frac{1}{2}, 3)$ on the line $y = 2 - 2x$?

21. Vicky decided to buy tickets to the local Razorbacks baseball games. She learned she must first become a Razorback Club member for $14 and then pay $2.50 per ticket to attend the games. Vicky purchased 12 games plus the membership fee. The ticket sales person charged her $34. Was she charged correctly? If not, how much should she have been charged? Use words and/or numbers to show how you determined your answer.

In each set of three linear equations, two are equivalent. Identify the one linear equation that is not equivalent to the others in the set. Show all work necessary to justify your answer.

22. $\begin{cases} 2x + 4y = 8 \\ y = \frac{1}{2}x - 2 \\ y = -\frac{1}{2}x + 2 \end{cases}$

23. $\begin{cases} y = 3x - 4 \\ y = 3(x - 2) + 2 \\ -3x + y = 4 \end{cases}$

24. $\begin{cases} y = x - 1 \\ 5x + 5y = -5 \\ y = -(x + 6) + 5 \end{cases}$

Lesson 3.5 ~ More Graphing Linear Equations

REVIEW

Write an equation in slope-intercept form that is represented by the given information.

25.

x	y
0	6
1	4
2	2
3	0

26. has a slope of –2 and y-intercept of 1

27.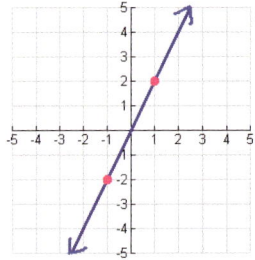

28. goes through the points (–3, 3) and (2, 8)

29.

30.

x	y
1	–1
3	9
6	24
8	34

31.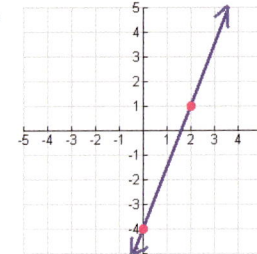

32.

x	y
4	–1
4	3
4	5
4	7

33. has a slope of $\frac{1}{3}$ and goes through the point (6, 1)

Tic-Tac-Toe ~ Careers Using Algebra

Linear equations are an essential part of Algebra I. There are many career choices where knowledge of algebra is crucial. Research at least two different careers that require the use of algebra. Write a 1-2 page report about these careers.

Include the following in your report for each career:
- Description of the career
- How the career includes the use of algebra
- How much schooling is required for the career

Tic-Tac-Toe ~ Exponential Equations

Exponential equations are used to predict growth and decay. Exponential equations have a start value and multiplication as the repeated operation. The start value is represented by the variable b and the constant multiplier is represented by the variable m in the equation $y = b \cdot m^x$.

Most growth and decay situations involve percentages. For example, a car may depreciate (decrease in value) at a rate of 12% per year. Or the number of bacteria might increase by 40% each day. In order to find the value of m with percentages, you must first start with 100% and then add or subtract the given percentage depending on whether it is increasing or decreasing in value. The percent must also be converted to a decimal.

Car depreciating by 12% = 100% − 12% = 88% = 0.88
Bacteria increasing by 40% = 100% + 40% = 140% = 1.40

Numbers to be used as m in the equation.

A car was purchased for $14,000 and depreciates at a rate of 12% per year. This means it keeps 88% of its value. The exponential equation would be:
$$y = 14{,}000 \cdot 0.88^x$$

You can find the new value of the car after any number of years by substituting the number of years the car has depreciated for x. For example, to find the value of the car after 5 years, substitute 5 for x. Remember to follow the order of operations when calculating.
$$y = 14{,}000 \cdot 0.88^5 \approx \$7{,}388.25$$

Answer each of the following exercises using the exponential equation $y = b \cdot m^x$.

1. Mary bought a brand new car in 2012 for $22,000. She was told that her model of car has an annual depreciation rate of 11%.
   a. How much was Mary's car worth if she sold it in 2015?
   b. How much will her car be worth if she waits and sells it in 2023?

2. Imar made $4,000 over the summer and wants to invest it for the future. He finds a bank that offers him 6% growth each year.
   a. How much will Imar have in the account after 3 years?
   b. How much will he have after 8 years?
   c. After approximately how many years will he have doubled his money?

3. The number of bacteria in a kitchen sink increases by 60% each day the sink remains uncleaned. The sink currently has 15 bacteria.
   a. How many bacteria will be present after 10 days of not washing the sink?
   b. Approximately how long will it take before there are at least 1,000 bacteria in the sink?

4. Write two of your own growth or decay application problems and find the solutions.

Lesson 3.5 ~ More Graphing Linear Equations

GRAPHING LINEAR INEQUALITIES IN TWO VARIABLES

LESSON 3.6

 Graph a linear inequality on a coordinate plane.

EXPLORE! IN THE SHADE

Step 1: Juniper is thinking of two numbers, x and y, whose sum is at least 6.
  a. Write an inequality that represents this statement.
  b. Find 6 pairs of numbers Juniper may be thinking of. List the sets of two numbers as ordered pairs (x, y).

Step 2: How many possible pairs of numbers fit Juniper's description? Explain your reasoning.

Step 3: On a coordinate plane, graph at least three ordered pairs that have a sum of exactly six (remember that Juniper said the sum of her numbers is at least six). Draw a solid line through these points. This is the boundary line for the solution set. All values on this line have ordered pairs that have a sum of 6.

Step 4: Plot at least three of your ordered pairs from **Step 1** on your graph. Are these numbers "above" or "below" your boundary line? Is it possible for some of your points to be on one side of the boundary line and other points on the opposite side? Explain your reasoning.

Step 5: Shade the coordinate plane on the side of the boundary line you placed your ordered pairs on in **Step 4**. This shows all of the possible solutions to Juniper's description of her two numbers. Choose another point in your shaded area as a test point. Do the x- and y-values add up to more than 6? If so, you shaded the correct side of the boundary line.

Step 6: On a new coordinate plane, begin to graph the inequality $y < \frac{1}{2}x - 1$ by graphing three points on the line $y = \frac{1}{2}x - 1$. DO NOT connect the points with a solid line. Test one of the points on your line in your inequality. What do you notice?

Step 7: Connect your points with a dashed line to show that these points are not included in the solution. Choose a test point on either side of the boundary line. Substitute the point into the original inequality. If the x- and y-values make the statement true, shade that side of the boundary line. If they do not, shade the side that does not contain your test point.

Step 8: Graph $y \leq -2x + 3$ on a new coordinate plane.

GRAPHING LINEAR INEQUALITIES IN TWO VARIABLES

1. Graph the boundary line by replacing the inequality sign with an equal sign.
    a. If the inequality sign is > or <, use a dashed boundary line.
    b. If the inequality sign is ≥ or ≤, use a solid boundary line.
2. Choose a test point which is not on the boundary line. Substitute the *x*-and *y*-values into the original inequality.
3. If the test point makes a true statement when substituted into the inequality, shade the side of the boundary line that contains that point. If the statement is false, shade the side of the boundary line that does not contain the test point.

EXAMPLE 1 Graph $y > 3x - 4$.

SOLUTION

Graph the boundary line by graphing $y = 3x - 4$. Use a dashed line to connect the points since the inequality symbol is >.

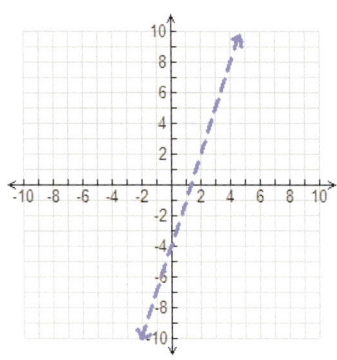

Choose a test point that is not on the boundary line. $(-2, 4)$

Substitute the coordinates into the original inequality to see if the statement is true or false. $4 \overset{?}{>} 3(-2) - 4$

Four is greater than negative 10 so the statement is true. $4 > -10$

Shade the entire side of the boundary line that contains the test point $(-2, 4)$.

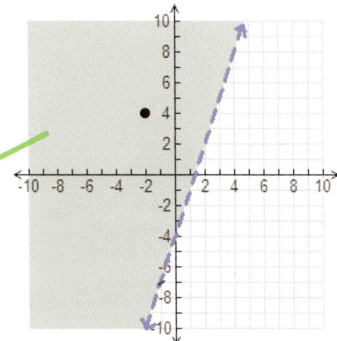

The shaded region shows all the possible ordered pairs that are solutions to $y > 3x - 4$.

Lesson 3.6 ~ Graphing Linear Inequalities in Two Variables

EXAMPLE 2

Graph $x + 2y \leq 4$.

SOLUTION

Rewrite the inequality as an equation.
Convert to slope-intercept form.

$$x + 2y = 4$$
$$\underline{-x \qquad\quad -x}$$
$$2y = -x + 4$$
$$y = -\frac{1}{2}x + 2$$

Graph the boundary line by graphing $y = -\frac{1}{2}x + 2$.
Use a solid line to connect the points since the inequality symbol is $\leq$.

Choose a test point that is not on the boundary line.   (4, 8)

Substitute the coordinates into the original inequality to see if the statement is true or false.

$$4 + 2(8) \stackrel{?}{\leq} 4$$

Twenty is NOT less than or equal to 4 so the statement is FALSE.

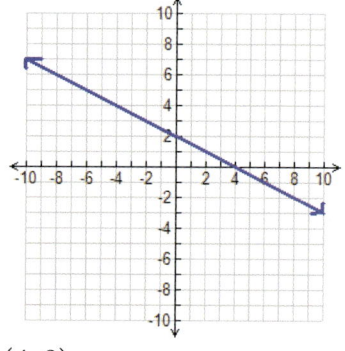

Shade the entire side of the boundary line that DOES NOT contain the test point (4, 8).

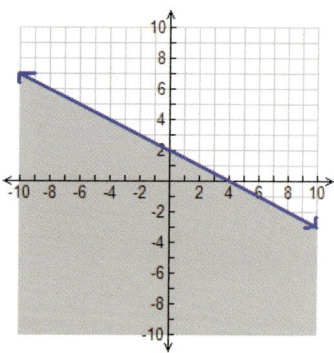

EXERCISES

1. Write a linear inequality in two variables that would use a dashed boundary line. Explain how you know the line would be dashed.

2. How are the graphs of linear inequalities similar to the graphs of linear equations? How are they different?

3. Fred is convinced that a solid line does not include the points on the line as part of the solution. Clint says the points on the line are part of the solution. Who is correct? Explain your reasoning.

4. How does a test point help you determine which side of the line to shade?

Graph each linear inequality.

5. $y > x + 1$

6. $y \leq \frac{2}{3}x - 5$

7. $y < -2x$

8. $y > -\frac{4}{3}x + 2$

9. $y \geq 4x - 7$

10. $y \leq -1 + \frac{3}{2}x$

11. $y < -\frac{2}{5}x + 2$

12. $x \leq -1$

13. $y > 3$

Match each linear inequality with its graph.

14. $y > x$

15. $y < x$

16. $y \geq x$

17. $y \leq x$

A.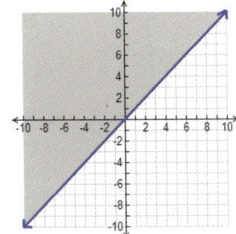
B.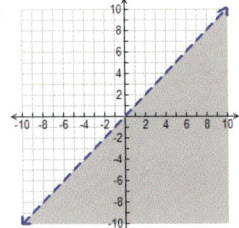
C.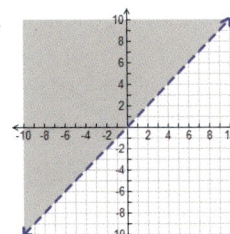
D.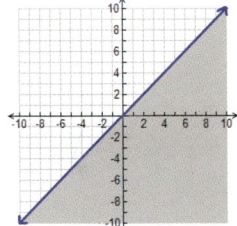

Graph each linear inequality.

18. $y + 3 > \frac{1}{3}(x + 6) - 1$

19. $2x - 5y \leq 10$

20. $-1 + y < 2(x + 1)$

21. Write the inequality for the graph below.

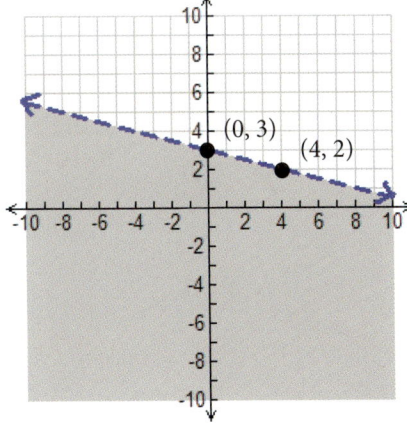

22. Write the inequality for the graph below.

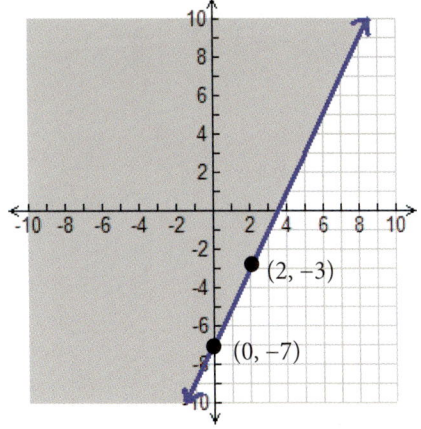

23. William is thinking of two numbers, x and y, whose sum is more than 2. Graph the solution set for his possible pairs of values.

Lesson 3.6 ~ Graphing Linear Inequalities in Two Variables

24. Ahn graphed the linear inequality $y > x + 5$. First, he graphed the line $y = x + 5$ by using a dashed line. He chose the test point (2, 8) which was above the line. He substituted (2, 8) into the equation he graphed for x and y.

$$8 \stackrel{?}{=} 2 + 5$$
$$8 \neq 7 \text{ FALSE}$$

Since the equation was false, he shaded below the line.
    a. What did Ahn do wrong?
    b. Draw the correct graph for the inequality.

25. Explain why a point on the boundary line should not be chosen as a test point for shading.

26. Taxis in Vickersville cost $1 per minute ($y$) plus $0.50 per mile ($x$). Janice has up to $30 to spend on a taxi ride.
    a. Write an inequality to represent this situation.
    b. Graph the inequality.
    c. Which quadrant(s) of your graph make sense for the situation? Explain your reasoning.

REVIEW

Write the linear equation for each of the given models.

27.

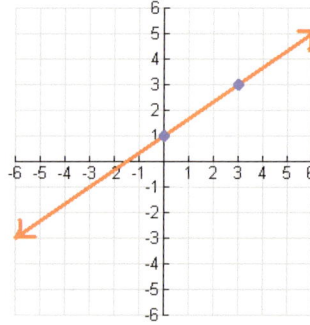

28.

x	y
1	4
2	7
3	10
4	13
5	16

29. Spot weighs 3 pounds. He gains two pounds every month. Let y represent Spot's weight and x represent the number of months.

30.

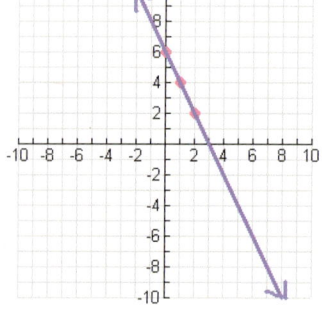

124   *Lesson 3.6 ~ Graphing Linear Inequalities in Two Variables*

INTRODUCTION TO NON-LINEAR FUNCTIONS

LESSON 3.7

 Recognize linear, quadratic, exponential and inverse variation functions.

In this textbook you have learned how to graph linear functions when written in different forms. However, not all functions produce a linear graph. In this lesson you will learn about a few types of **non-linear functions**. Non-linear functions are functions that, when graphed, do not form a line. There are a large variety of non-linear functions, many of which you will learn about in higher-level mathematics.

In a linear function, for each equal "step" that the *x*-value increases, the *y*-value increases or decreases a consistent amount.

$y = 4x + 5$

x	y
0	5
1	9
2	13
3	17
4	21

+4
+4
+4

In a non-linear function, there is not a consistent adding or subtracting pattern for each equal "step", although each type of non-linear function does have a unique pattern. In this lesson you will examine quadratic functions, exponential functions and inverse variation functions. Each type of non-linear function has a parent graph. The **parent graphs** are the most basic graph of each non-linear function.

EXPLORE!   NON-LINEAR CURVES

The Quadratic Function

Step 1: The parent graph of the quadratic function is the graph of $y = x^2$. Copy the table at right on your own paper and fill in the missing values.

Step 2: Plot the (*x*, *y*) points from the table. Connect the points with a curved line. Describe what the graph looks like.

$y = x^2$

x	y
−3	$(-3)^2 = 9$
−2	$(-2)^2 = 4$
−1	
0	
1	
2	
3	

The Exponential Function

Step 3: One parent graph of an exponential function is the graph of $y = 2^x$. Copy the table at right on your own paper and fill in the missing values using a calculator.

Step 4: Plot the (*x*, *y*) points from the table. Connect the points with a curved line. Describe what the graph looks like.

$y = 2^x$

x	y
−3	$2^{(-3)} = 0.125$
−2	$2^{(-2)} = 0.25$
−1	$2^{(-1)} = 0.5$
0	
1	
2	
3	

Lesson 3.7 ~ Introduction to Non-Linear Functions

The parent graphs of three non-linear functions are shown below. Other graphs in each "family" have the same shape as the parent graph but may be stretched, shrunk, moved or flipped.

Quadratic Functions

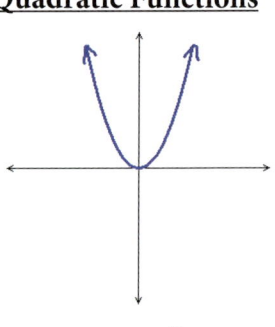

$y = x^2$

Exponential Functions

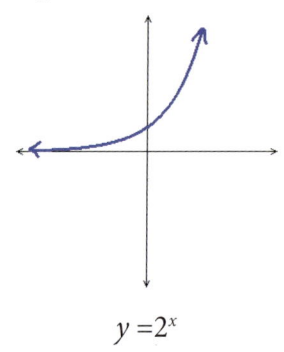

$y = 2^x$

Inverse Variation Functions

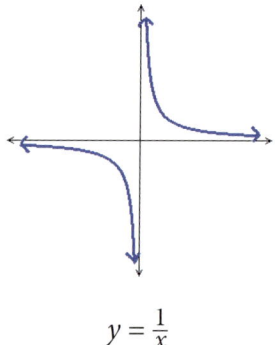

$y = \frac{1}{x}$

EXAMPLE 1

Determine if each graph, table or equation is linear or non-linear. If it is non-linear, identify the type of function (quadratic, exponential or inverse variation).

a. $y = 3x - 2$

b.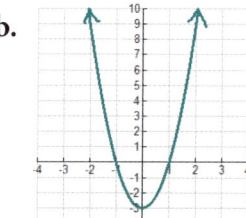

c.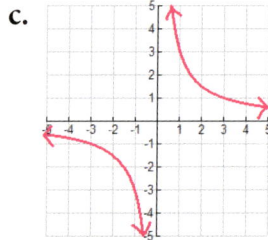

d.

x	y
−1	$\frac{1}{3}$
0	1
1	3
2	9
3	27

SOLUTIONS

a. LINEAR. This equation is linear because it is in slope-intercept form.

b. NON-LINEAR. This graph forms a "U" shape so it is a quadratic function.

c. NON-LINEAR. The graph matches the inverse variation parent graph.

d. NON-LINEAR. Graph the data points to see that it matches the parent function graph of the exponential function.

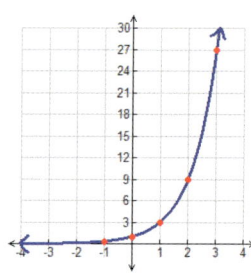

Lesson 3.7 ~ Introduction to Non-Linear Functions

Non-linear equations are used in the real world. Banks use exponential functions to calculate interest. Architects use quadratic functions when designing bridges. Businessmen use quadratic functions to determine costs of products in order to maximize profits. Chemists use exponential functions to calculate the rate of bacteria growth. An example of a real-world use of the inverse variation function is found in the creation of levers.

EXERCISES

Copy each table and use the given equation to fill in the missing values.

1. $y = x^2 + 2$

x	y
−2	$(-2)^2 + 2 = 6$
−1	
0	
1	
2	

2. $y = 3^x$

x	y
−2	$3^{(-2)} = \frac{1}{9}$
−1	
0	
1	
2	

3. $y = 2x - 5$

x	y
−2	$2(-2) - 5 = -9$
−1	
0	
1	
2	

4. Graph the five points from the table in **Exercise 1** on a coordinate plane. What type of non-linear equation is this?

5. Graph the five points from the table in **Exercise 2** on a coordinate plane. What type of non-linear equation is this?

6. Graph the five points from the table in **Exercise 3** on a coordinate plane. What type of equation is this?

Determine if each graph, table or equation is linear or non-linear. If it is non-linear, identify the type of graph (quadratic, exponential or inverse variation).

7.

x	y
−2	$\frac{1}{16}$
−1	$\frac{1}{4}$
0	1
1	4
2	16

8. $y = (x + 1)^2$

9.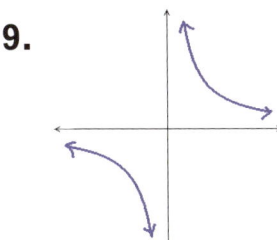

10.

11.

x	y
−2	−2
−1	1
0	4
1	7
2	10

12. $y = 4^x$

Lesson 3.7 ~ Introduction to Non-Linear Functions

13. A bacteria culture begins with 5 cells and doubles every hour. This situation can be represented by the exponential function $y = 5 \cdot 2^x$ where x represents the number of hours and y represents the number of bacteria. How many bacteria will be in the culture after 7 hours? Show all work necessary to justify your answer.

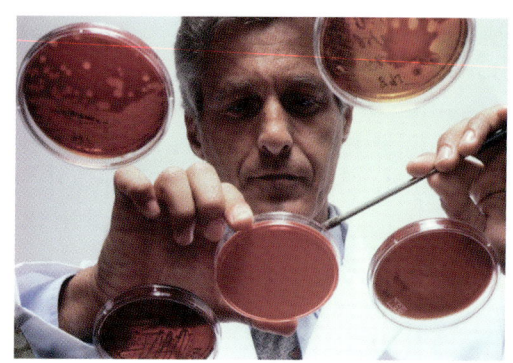

14. The path of a ball thrown through the air can be modeled by the equation $y = -4.9x^2 + 17x + 3.4$ when x is the time in seconds and y is the height of the ball in meters. Find the height of the ball after 3 seconds. Use words and/or numbers to show how you determined your answer.

15. How can you determine if a graph is linear or non-linear based on its equation?

16. Kyle deposits $1,000 in a savings account. Each year the bank gives him 4% interest based on the current value of the account.
   a. Create a table of values showing the value of his bank account after Years 1, 2 and 3.
   b. Is this relationship linear or non-linear? Explain your reasoning.

17. An arch was constructed as a memorial to those who had served in a recent war. The shape of the arch resembles a quadratic function with the equation $y = -\frac{1}{2}x^2 + 10x - 3$ where y is the height of the arch in meters and x is the distance from a fixed point. How high was the arch 14 meters from the fixed point? Use mathematics to justify your answer.

18. What type of graph (quadratic, exponential or inverse variation) is most likely to represent the population of a town that has many job opportunities? Explain your reasoning.

REVIEW

Solve each equation. Check your solution.

19. $x - 24 = 72$

20. $\frac{x}{5} = -11$

21. $8x + 13 = 61$

22. $17 = -4x + 2$

23. $2x + 7 = 5x - 8$

24. $3(x - 5) = -9$

Graph each equation on a coordinate plane. Clearly mark three points.

25. $y = \frac{1}{3}x - 3$

26. $y = 1$

27. $y = 2x + 1$

28. $y = x$

29. $y = -\frac{5}{2}x + 4$

30. $y = -4x + 9$

31. $x = -1$

32. $y = -\frac{1}{2}(x - 6) - 1$

33. $2x + 5y = 10$

Tic-Tac-Toe ~ Quadratic Functions

Quadratic functions can be described as being "U" shaped. Linear equations can be easily graphed when in slope-intercept form. Similarly, quadratic functions can be easily graphed when in factored form: $y = (x - a)(x - b)$. You need to know the two *x*-intercepts and the vertex (maximum or minimum point on the graph) in order to graph a simple quadratic function.

For example: Graph $y = (x - 4)(x + 2)$.

Step 1: Locate the *x*-intercepts. The *x*-intercepts can be found by setting the expressions inside each parentheses equal to 0 and solving.

$$x - 4 = 0 \qquad x + 2 = 0$$
$$+4 \quad +4 \qquad -2 \quad -2$$
$$x = 4 \qquad x = -2$$

Step 2: Average the *x*-intercepts by adding them together and then dividing the sum by 2. This is the *x*-coordinate of the vertex.

$$\frac{4 + -2}{2} = 1$$

Step 3: Substitute the number from **Step 2** back into the original equation to find the *y*-coordinate of the vertex.
$$y = (1 - 4)(1 + 2)$$
$$y = (-3)(3) = -9$$

Step 4: Graph the quadratic function. Connect the three points with a smooth curve.

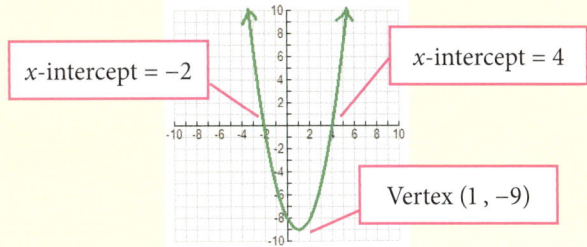

Graph the following quadratic functions following the four steps shown above.

1. $y = (x - 2)(x - 6)$

2. $y = (x - 5)(x + 1)$

3. $y = (x - 2)(x + 2)$

4. $y = (x + 5)(x + 1)$

REVIEW BLOCK 3

Vocabulary

continuous equivalent parent graph
discrete   non-linear functions slope-intercept form

- Graph linear equations in slope-intercept form.
- Write a linear equation for a given graph.
- Write a linear equation in slope-intercept form when given information about the line.
- Convert different forms of linear equations to slope-intercept form.
- Graph linear equations that are not written in slope-intercept form.
- Graph a linear inequality on a coordinate plane.
- Recognize linear, quadratic, exponential and inverse variation functions.

Lesson 3.1 ~ Graphing Using Slope-Intercept Form

Draw a coordinate plane for each problem and graph the given equation. Clearly mark three points on the line.

1. $y = 3x - 4$

2. $y = \frac{2}{3}x + 3$

3. $y = x - 1$

4. $y = -2x$

5. $x = 3$

6. $y = 6 + \frac{4}{3}x$

7. Create a linear equation that satisfies each condition. Graph your equations on a coordinate plane.
  a. Slope = 2 and a negative y-intercept
  b. Slope = 0 and a y-intercept of -3
  c. A negative slope and a positive y-intercept
  d. A positive slope and a y-intercept of 0

Lesson 3.2 ~ Writing Linear Equations for Graphs

Identify the slope and y-intercept of each graph and write the corresponding linear equation in slope-intercept form.

8.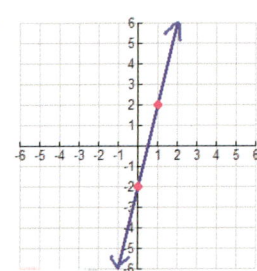

9.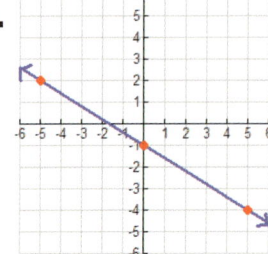

10.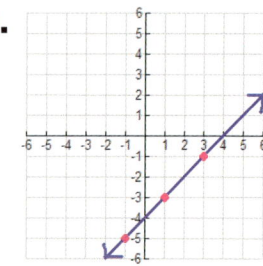

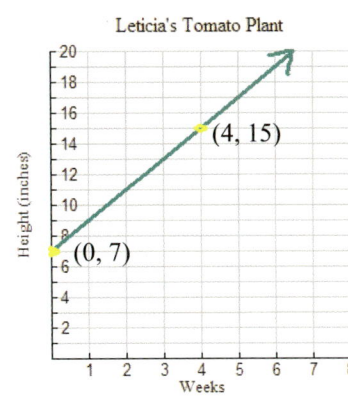

11. At two different times during the summer, Leticia measured the height of a tomato plant she had planted in June. She measured it when she first planted it and then again 4 weeks later.

a. Find the slope-intercept equation that represents the height of Leticia's tomato plant based on the number of weeks since she planted it.

b. Use your equation to determine exactly how tall the tomato plant will be after 7 weeks.

c. Determine how many weeks have passed if the plant is 29 inches tall. Use mathematics to justify your answer.

Lesson 3.3 ~ Writing Linear Equations from Key Information

Write an equation in slope-intercept form when given key information about a line.

12. slope = $\frac{3}{4}$, y-intercept = 5

13. slope = −5, y-intercept = 1

14. slope = 1, y-intercept = 9

15. slope = $\frac{2}{5}$, y-intercept = 0

16. slope = 2, goes through the point (2, 3)

17. slope = $\frac{1}{2}$, goes through the point (6, 1)

18. slope = −1, goes through the point (−3, 5)

19. slope = $\frac{5}{2}$, goes through the point (−6, −10)

20. goes through the points (1, 1) and (5, 9)

21. goes through the points (−6, 0) and (3, 3)

22. goes through the points (−4, 8) and (−3, 5)

23. goes through the points (8, −4) and (5, −4)

24. A furniture rental company rents large screen televisions. They charge an initial fee plus $20 for each day the TV is rented. Steven rented a TV for 8 days and was charged $225. Let x represent the number of days and y represent the total cost of the rental.

a. Identify the slope and one ordered pair from the information given.

b. Find the equation of the line that fits this information.

c. If another customer rents a TV for 17 days, how much should he expect to pay?

25. A canoe rental company on Deep Sea Lake rents canoes for a set fee plus an additional charge per hour. Marshall asked two different individuals how many hours they had rented their canoes for and how much it cost. One rented a canoe for 4 hours and paid $32. Another person rented a canoe for 10 hours for $56. Let x represent the length of time in hours and let y represent the total cost.

a. What is the linear equation that represents the data?

b. What number in the linear equation represents the amount of the set fee?

c. What is the real-world meaning of the slope in this equation?

d. How much will someone pay for a canoe rental from this company if he keeps the canoe for 6 hours? Show all work necessary to justify your answer.

Lesson 3.4 ~ Different Forms of Linear Equations

Convert each equation to slope-intercept form.

26. $y = 4 + 2(x - 7)$

27. $3x + 6y = 18$

28. $y = \frac{1}{4}(x + 4) - 3$

29. $4x - 5y = 15$

30. $-x + 3y = -12$

31. $y - 2 = 3(x + 1)$

Lesson 3.5 ~ More Graphing Linear Equations

Convert each equation to slope-intercept form and graph. Clearly mark at least three points on each line.

32. $-4x + 2y = -6$

33. $y + 1 = 3(x - 2)$

34. $y = \frac{3}{2}(x - 4) + 2$

35. $7x = -14$

36. $x + 3y = 12$

37. $y = 2(x - 1) + 2$

Determine if each point is on the given line. Show all work necessary to justify your answer.

38. Is the point $(-2, 1)$ on the line $4x - 3y = -11$?

39. Is the point $(2, 5)$ on the line $y = 2(x - 1) + 3$?

40. Is the point $(-6, 0)$ on the line $y + 4 = \frac{1}{2}(x + 4) + 3$?

Lesson 3.6 ~ Graphing Linear Inequalities in Two Variables

Graph each linear inequality.

41. $y < 2x - 4$

42. $y \geq \frac{3}{4}x - 2$

43. $y < -x + 1$

44. $y > -\frac{1}{2}x$

45. $y \geq -4$

46. $2x + 3y < 9$

Match each linear inequality with its graph.

47. $y \geq \frac{1}{2}x - 3$

48. $y \leq \frac{1}{2}x - 3$

49. $y < \frac{1}{2}x - 3$

50. $y > \frac{1}{2}x - 3$

A.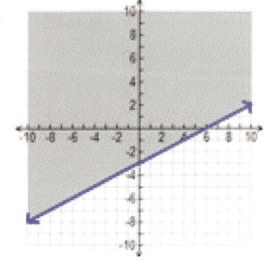
B.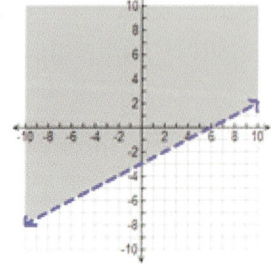
C.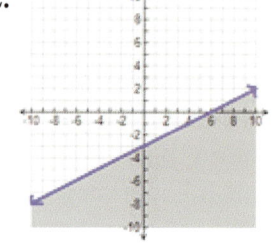
D.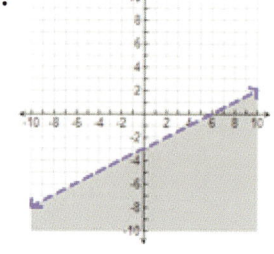

Lesson 3.7 ~ Introduction to Non-Linear Functions

Determine if each graph, table or equation is linear or non-linear. If it is non-linear, identify the type of graph (quadratic, exponential or inverse variation).

51.

x	y
−2	12
−1	3
0	0
1	3
2	12

52. $y = \frac{2}{3}x - 4$

53.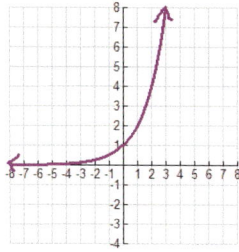

54.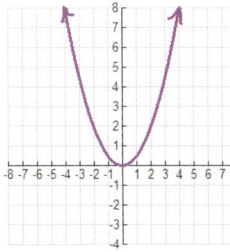

55.

x	y
−2	0
−1	2
0	4
1	6
2	8

56.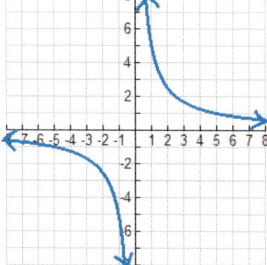

57. Nicki deposits $4,000 in a savings account. Each year the bank gives her 3% interest based on the current value of the account.
  a. Create a table of values showing the value of her bank account after Years 1, 2 and 3.
  b. Is this relationship linear or non-linear? Explain your reasoning.

Tic-Tac-Toe ~ Graphing Design

Lines are used in many types of artwork. Use a large sheet of graph paper to create a piece of artwork.

Step 1: Draw a coordinate plane that includes all four quadrants.

Step 2: Create a design using at least 15 different lines. Make sure over two-thirds of the lines are not vertical or horizontal.

Step 3: Write the equations for each line on the back of your piece of artwork.

Step 4: Color your artwork and sign the bottom right corner.

CAREER FOCUS

Scott
Retail Manager

I am a retail manager for a department store. My position plays an important part in making sure that our department makes money. Many decisions that I make affect how much money our department brings in. I try to maximize our profits by making sure that our products are good and are priced right. I also determine how many employees to hire and how many should work each day. If too many people are working, profits get spent on labor we do not need.

I use math every day in my job as a retail manager. I use basic math skills in my job as well as complex equations. Math helps me to determine how to get the most profits for the department. The profit of my department is affected by "shrink." Shrink is a term that we use in retail to describe theft, damaged or broken products, and products that we never received but still get billed for. Profits are also affected by the cost of transportation, electricity, building maintenance and advertising. As a retail manager, I can control how many employees we are using and also reduce shrink. Both of these things will improve the company's total profits.

I was hired as a retail manager after I had completed my college degree. A person can get hired in my career with a high school diploma, but will need to go through lots of on the job training. People usually have to work their way up through the company to get into a management position.

Salaries for retail managers can vary quite a bit. Salaries start as low as minimum wage for people without experience or other training. People who get into a management training program can earn around $25,000 per year. After becoming a manager, salaries range from $40,000 to $60,000 per year.

I like my job as a retail manager because I enjoy working with the public. I also like how fast things change in the retail world. The best part of my job, though, is helping new employees turn into great long-term members of our team.

CORE FOCUS ON LINEAR EQUATIONS
BLOCK 4 ~ SYSTEMS OF EQUATIONS

LESSON 4.1	PARALLEL, INTERSECTING OR THE SAME LINE	137
	EXPLORE! TYPES OF SYSTEMS	
LESSON 4.2	SOLVING SYSTEMS BY GRAPHING	141
LESSON 4.3	SOLVING SYSTEMS USING TABLES	145
	EXPLORE! LARRY'S LANDSCAPING	
LESSON 4.4	SOLVING SYSTEMS BY SUBSTITUTION	150
	EXPLORE! A TRIP ON I-70	
LESSON 4.5	SOLVING SYSTEMS USING ELIMINATION	154
LESSON 4.6	CHOOSING THE BEST METHOD	158
	EXPLORE! WHAT'S EASIEST?	
LESSON 4.7	APPLICATIONS OF SYSTEMS OF EQUATIONS	161
	EXPLORE! AT THE MOVIES	
LESSON 4.8	SYSTEMS OF LINEAR INEQUALITIES	167
LESSON 4.9	CONVERTING REPEATING DECIMALS TO FRACTIONS	171
REVIEW	BLOCK 4 ~ SYSTEMS OF EQUATIONS	175

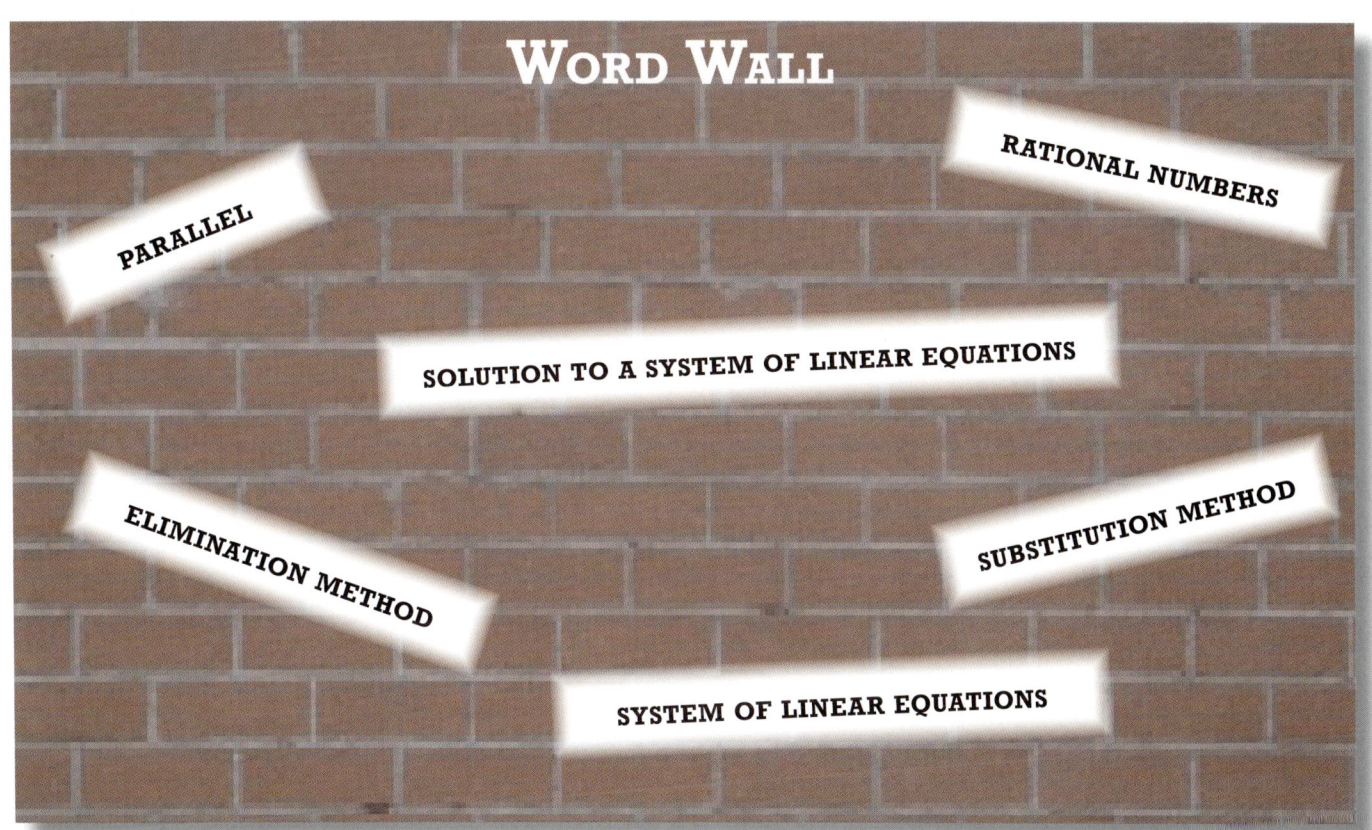

BLOCK 4 ~ SYSTEMS OF EQUATIONS
TIC-TAC-TOE

How Many Solutions?

Determine whether a system of three linear equations has zero, one or infinitely many solutions.

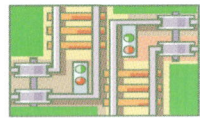

See page 140 for details.

Pros and Cons

Create a visual display showing the positive and negative aspects of each method for solving systems of equations.

See page 174 for details.

Polygons

Develop systems of equations that make different polygons when graphed.

See page 153 for details.

"How To" Guide

Produce a brochure about solving systems of linear equations using at least two different ways.

See page 174 for details.

Fraction Coefficients

Solve systems of equations with fraction coefficients.

See page 166 for details.

Math Dictionary

Make a dictionary for all the vocabulary terms in this textbook. Create diagrams when possible.

See page 149 for details.

Letter to Fifth Graders

Write a letter to a fifth grade class explaining why learning math is important. Support your reasons with research.

See page 153 for details.

Different Systems

Find the solutions to systems of equations involving a quadratic and a linear function.

See page 149 for details.

Solution Given

Create systems of equations that have given solutions.

See page 157 for details.

PARALLEL, INTERSECTING OR THE SAME LINE

LESSON 4.1

 Algebraically determine if two lines are parallel, intersecting or the same line.

In this block, you will look at two linear equations at the same time. A set of two or more linear equations that have common variables is called a **system of linear equations**. Systems of linear equations have three different types of solutions.

EXPLORE! — TYPES OF SYSTEMS

System #1
$y = \frac{1}{2}x - 2$
$6x + 2y = 10$

System #2
$y = -3 + 4x$
$y = 4(x + 1) - 7$

System #3
$y = -\frac{2}{3}x - 1$
$2x + 3y = 6$

Step 1: For each system, convert all equations into slope-intercept form.

Step 2: Draw three coordinate planes. Graph the two lines in System #1 on the first coordinate plane, graph System #2 on the second coordinate plane and System #3 on the third coordinate plane.

Step 3: Describe in words how the two lines in System #1 are related.

Step 4: Describe in words how the two lines in System #2 are related.

Step 5: Describe in words how the two lines in System #3 are related.

Step 6: Is there a way to tell, just by looking at the equations in slope-intercept form, when the lines will be intersecting, parallel or the same line? Explain your reasoning.

Step 7: Without graphing, how do you think the equations in System #4 are related? Explain your reasoning.

System #4
$y = 5x - 1$
$y = 5x + 4$

A **solution to a system of linear equations** is the ordered pair (x, y) that satisfies both linear equations in the system. The solution to the system of linear equations is found at the point of intersection of the two lines. Systems of linear equations can have one solution, no solutions or infinitely many solutions.

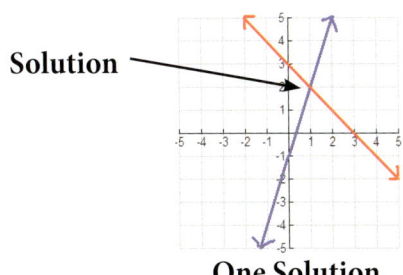

Solution → **One Solution** Intersecting Lines

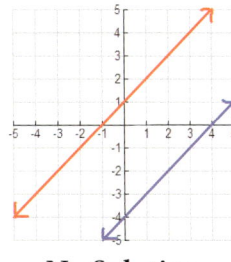

No Solution Parallel Lines

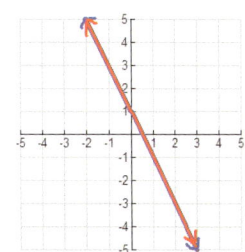

Infinitely Many Solutions Same Line

Lesson 4.1 ~ Parallel, Intersecting or the Same Line **137**

A set of two lines that intersects has one solution. A set of two lines that are **parallel** has no solutions because parallel lines never intersect. If a system of linear equations contains two equations that represent the same line, there are infinitely many solutions because the two lines intersect at an infinite number of points.

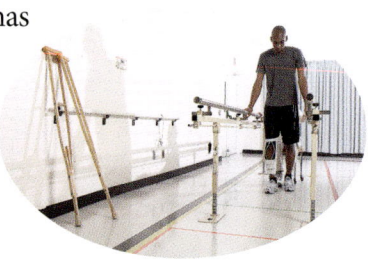

If the linear equations are in slope-intercept form, it is possible to determine whether the two lines are intersecting, parallel or the same lines just by looking at their slopes and y-intercepts.

> **DETERMINING TYPES OF SYSTEMS**
> 1. If the two lines have the SAME slope and SAME y-intercept, the two lines are the same line.
> 2. If the two lines have the SAME slope but DIFFERENT y-intercepts, the lines are parallel.
> 3. If the two lines have DIFFERENT slopes, the lines are intersecting.

EXAMPLE 1 Determine if the two lines in each system of equations are intersecting, parallel or the same line. State how many solutions there will be for each system.
a. $4x - 5y = 30$
   $y = \frac{4}{5}x - 3$
b. $-x + 2y = -6$
   $y = \frac{1}{2}(x + 2) - 4$
c. $y = 2x + 1$
   $y = -2x + 1$

SOLUTIONS a. Convert equations to slope-intercept form.

$4x - 5y = 30$
$-4x \qquad -4x$

$\frac{-5y}{-5} = \frac{30 - 4x}{-5}$

$y = -6 + \frac{4}{5}x$ or

$\boxed{y = \frac{4}{5}x - 6}$ $\qquad \boxed{y = \frac{4}{5}x - 3}$

Compare the two equations. The slopes of the two lines are the same but the y-intercepts are different. This means that the two lines are <u>parallel</u>. There are <u>no solutions</u> to this system.

b. Convert both equations to slope-intercept form.

$-x + 2y = -6$
$+x \qquad +x$

$\frac{2y}{2} = \frac{x - 6}{2}$ $\qquad y = \frac{1}{2}(x + 2) - 4$

$\qquad\qquad\qquad\qquad y = \frac{1}{2}x + 1 - 4$

$\boxed{y = \frac{1}{2}x - 3}$ $\qquad \boxed{y = \frac{1}{2}x - 3}$

Compare the two equations in slope-intercept form, $y = \frac{1}{2}x - 3$ and $y = \frac{1}{2}x - 3$. The slopes and y-intercepts of the two lines are the same. This means that the two lines are the <u>same line</u> and there are <u>infinitely many solutions</u> to this system.

c. Both equations are already in slope-intercept form. The equations have different slopes, therefore the two lines are <u>intersecting</u>. There is <u>one solution</u>.

Lesson 4.1 ~ Parallel, Intersecting or the Same Line

EXERCISES

Determine if each graph shows a system of linear equations that is intersecting, parallel or the same line. State how many solutions there are for each system.

1.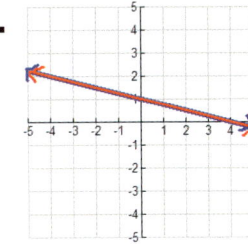

2.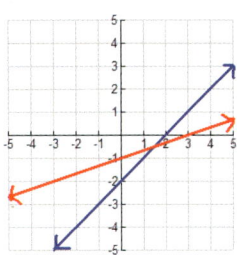

3.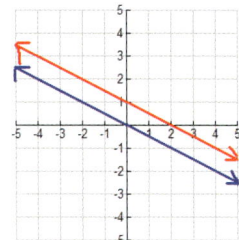

Graph the two linear equations in each system on a single coordinate plane. State whether the lines are intersecting, parallel or the same line.

4. $y = -3x$
   $y = \frac{1}{3}x + 3$

5. $y = 3 + \frac{1}{2}x$
   $y = \frac{1}{2}x + 3$

6. $y = 1 + 4(x - 1)$
   $y = 4(x + 1) + 1$

Determine if the two lines in each system of equations are intersecting, parallel or the same line. State how many solutions there will be for each system. Use words and/or numbers to show how you determined your answer.

7. $y = -3x + 4$
   $y = -3x + 3$

8. $y = -\frac{2}{3}x + 5$
   $y = 3x - 5$

9. $2y = -4x + 12$
   $y = 2x + 4$

10. $y = \frac{1}{3}x - 2$
    $-3x + 6y = -12$

11. $6y = 15$
    $10y = 25$

12. $6x + y = 5$
    $y = 6x - 3$

13. Two ants crawled across a piece of graph paper. One followed the path of the linear equation $3x + 2y = 8$. The other ant followed the path of the linear equation $2x + 3y = 6$. Will the ants' paths cross? How do you know?

14. A parallelogram was formed by the intersection of the four lines whose equations are given. Determine algebraically which pairs of sides ($\overline{AB}, \overline{BC}, \overline{CD}, \overline{DA}$) are parallel. Verify your answer by graphing.

$\overline{AD}: 2x + 3y = 6$
$\overline{AB}: -6x + 4y = 0$
$\overline{BC}: y = -\frac{2}{3}x + 4$
$\overline{CD}: y = -5 + \frac{3}{2}x$

15. Kirk and Samantha walk home from school. The map of their town was placed on a coordinate grid. Kirk walked home following the linear equation $5x + 4y = 28$. Samantha walked home following the path of $y = -\frac{5}{4}x + 7$. Describe the similarities or differences in their paths home.

16. Write a system of two linear equations in which the lines will intersect. Graph the two lines on the same coordinate plane. Use words, graphs and/or numbers to justify your answer.

17. Write a system of two linear equations in which the lines are parallel. Graph the two lines on the same coordinate plane. Use words, graphs and/or numbers to justify your answer.

18. Describe how you can tell if two lines intersect by looking at the linear equations in slope-intercept form.

19. On her Block 4 Test, Victoria was asked to give an example of two lines that are parallel but not the same line. She answered with the equations: $y = 4x + 5$ and $y = 3x + 5$. Did she get the question right? If not, what mistake did she make?

REVIEW

State whether each equation is true or false for the values of the variables given. Show all work necessary to justify your answer.

20. $5x + 2y = 10$ where $x = 0$ and $y = 5$

21. $-3x + y = 7$ where $x = -1$ and $y = -4$

22. $y = \frac{4}{3}x - 2$ where $x = 9$ and $y = 34$

23. $y = 1 + 2(x - 5)$ where $x = 7$ and $y = 5$

Simplify each expression.

24. $4 + 6x - 1 + 2x$

25. $3(x - 2) + 2(x + 7)$

26. $5x + x + 7x - 10x$

27. $6(x - 1) - 2(x + 1)$

28. $7x + 3y - x + 4y - 2x$

29. $3(2x + 4y) + 5(x - 2y)$

TIC-TAC-TOE ~ HOW MANY SOLUTIONS?

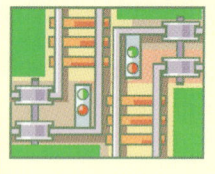

In this block, all the systems of linear equations only include two equations; however, systems of equations can include more than two equations. A solution to a system of linear equations is the point where all the lines intersect. Each of the systems below has either zero, one or infinitely many solutions. Use input-output tables or graphing to determine the number of solutions. If the system does have one solution, give the point of intersection.

System #1
$y = 3x + 3$
$y = 3x - 4$
$y = 3x - 7$

System #2
$y = 2(x - 3) + 5$
$4x - 2y = 2$
$y = 2x - 1$

System #3
$y = \frac{1}{2}x - 3$
$y = x - 5$
$y = -\frac{3}{4}x + 2$

System #4
$y = \frac{1}{2}x$
$-x + 2y = 6$
$y = \frac{1}{2}(x + 4) - 1$

System #5
$y + x = 6$
$2x + y = 8$
$-x + y = 2$

SOLVING SYSTEMS BY GRAPHING

LESSON 4.2

 Determine the solution to a system of equations by graphing.

Systems of linear equations can have zero, one or infinitely many solutions. A solution to a system of linear equations is the ordered pair (x, y) that satisfies both linear equations in the system. The solution is the point(s) where the two lines intersect. The solution is stated by giving the coordinates for the point(s) where the two lines intersect.

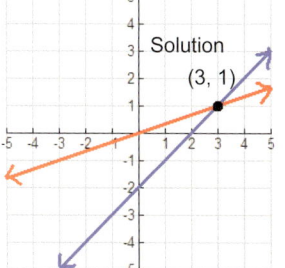

✓ It is important to always check your solution by verifying that the ordered pair (x, y) creates a true statement when substituted into each linear equation in the system.

EXAMPLE 1

Sue graphed the following systems. She listed her solution to each system below. Decide whether her ordered pair is a solution to the system of equations.

a. $-2x + 4y = 20$
$3x + y = -9$
Sue's Answer: $(-4, 3)$

b. $y = \frac{1}{3}x + 3$
$x + y = 9$
Sue's Answer: $(-9, 0)$

SOLUTIONS

a. Substitute -4 for x and 3 for y in each linear equation in the system.

$$-2x + 4y = 20 \qquad\qquad 3x + y = -9$$
$$-2(-4) + 4(3) \stackrel{?}{=} 20 \qquad\qquad 3(-4) + 3 \stackrel{?}{=} -9$$
$$8 + 12 \stackrel{?}{=} 20 \qquad\qquad -12 + 3 \stackrel{?}{=} -9$$
$$20 = 20 \qquad\qquad -9 = -9$$

The ordered pair makes each equation true. It is a solution to the system of linear equations.

b. Substitute -9 for x and 0 for y in each linear equation in the system.

$$y = \frac{1}{3}x + 3 \qquad\qquad x + y = 9$$
$$0 \stackrel{?}{=} \frac{1}{3}(-9) + 3 \qquad\qquad -9 + 0 \stackrel{?}{=} 9$$
$$0 \stackrel{?}{=} -3 + 3 \qquad\qquad -9 \neq 9$$
$$0 = 0$$

The ordered pair works in the first equation but does not work in the second equation. It is NOT the solution to the system of linear equations.

SOLVING SYSTEMS OF LINEAR EQUATIONS BY GRAPHING

1. Convert both linear equations in the system to slope-intercept form.
2. Graph both equations on the same coordinate plane. Be sure to clearly mark at least three points on each line.
3. Determine the point of intersection.
4. Verify that the ordered pair is the solution by substituting the *x*-value and *y*-value into each equation in the system.

EXAMPLE 2

Solve the system of equations by graphing. Check the solution.
$$y = \tfrac{1}{2}x - 3 \quad \text{and} \quad 3x + 2y = 2$$

SOLUTION

Convert both equations to slope-intercept form.

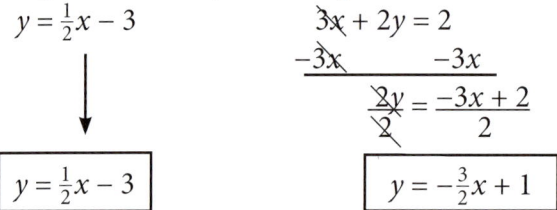

Graph both equations on the same coordinate plane. Start at the *y*-intercept and create at least two more points on each line before drawing the lines.

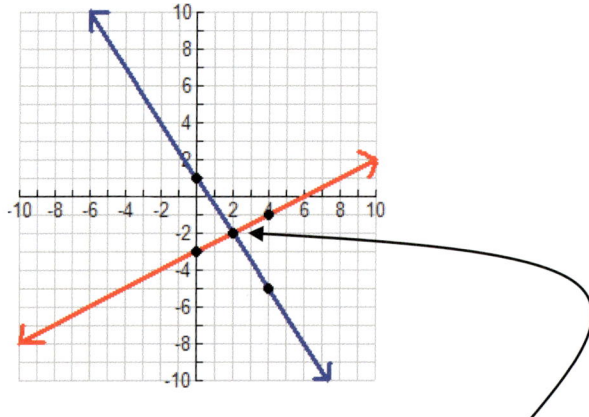

Determine the point of intersection of the two lines → (2, −2).

☑ Check the solution (2, −2) by substituting the 2 in for *x* and −2 in for *y* in each equation to determine if they make the equation true.

$$y = \tfrac{1}{2}x - 3 \qquad\qquad 3x + 2y = 2$$
$$-2 \stackrel{?}{=} \tfrac{1}{2}(2) - 3 \qquad 3(2) + 2(-2) \stackrel{?}{=} 2$$
$$-2 \stackrel{?}{=} 1 - 3 \qquad\qquad 6 + -4 \stackrel{?}{=} 2$$
$$-2 = -2 \qquad\qquad\qquad 2 = 2$$

The point (2, −2) is the solution to the system.

EXERCISES

1. Al was sick when his math class learned how to solve systems of linear equations by graphing. Explain the process Al needs to follow to find the solution to a system of linear equations by graphing.

Use each graph to solve the system of linear equations.

2.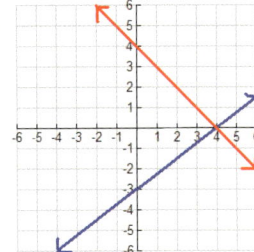

3.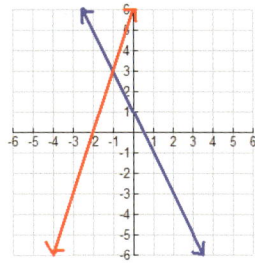

4.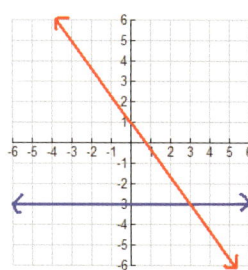

Decide whether the given ordered pair is a solution to the system of equations. Show all work necessary to justify your answer.

5. $y = -4x$
   $y = \frac{1}{2}x + 7$
   $(-2, 8)$

6. $x + 5y = 8$
   $4x - 5y = 7$
   $(3, 1)$

7. $y = x - 9$
   $x + y = -1$
   $(4, -5)$

8. Polly took a quiz on solving systems of linear equations. She was not sure how she did on two of the harder problems when she turned in the quiz. Later, her friend told her she should have checked her answers. Did Polly answer the questions correctly? Use words and/or numbers to support your answer.

   a. $y = 4x - 1$
      $2x - y = -13$
      Polly's solution $(-2, -9)$

   b. $3x + 6y = 15$
      $-2x + 3y = -3$
      Polly's solution $(3, 1)$

Solve each system of equations by graphing. Show all work necessary to prove that your answer is correct. If there is not exactly one solution, determine if the system of equations has infinitely many solutions or no solution.

9. $y = -\frac{1}{4}x + 6$
   $y = \frac{1}{2}x + 3$

10. $y = x - 2$
    $y = -2x + 1$

11. $y = \frac{1}{2}x + 1$
    $y = 4$

12. $y = -\frac{1}{4}x + 1$
    $x + 4y = 4$

13. $y = \frac{1}{3}x + 2$
    $y = -\frac{2}{3}x + 5$

14. $y = \frac{2}{3}x - 3$
    $x = -3$

15. $y = -5 + \frac{3}{2}x$
    $-2x + 4y = 4$

16. $y = -3x + 4$
    $2x - y = 1$

17. $2x + 5y = 5$
    $y = -\frac{2}{5}x + 4$

Lesson 4.2 ~ Solving Systems by Graphing

18. Sarah begins the year with $100 in her savings account. Each week, she spends $8. Martin begins the year with no money saved, but each week he puts $12 into an account. Let *x* represent the number of weeks since the beginning of the year and *y* represent the total money in the account.
   a. Write a linear equation to represent Sarah's total money in her savings based on the number of weeks that have passed.
   b. Write a linear equation to represent Martin's total money in his savings based on the number of weeks that have passed.
   c. Graph both equations on the same first quadrant coordinate plane.
   d. At what point do the lines intersect? What is the real-world meaning of this point?

19. The perimeter of Karen's rectangular garden is 42 feet. The length of the garden is 3 feet more than twice the width. Let *y* represent the length of the garden and *x* represent the width of the garden.
   a. Write a linear equation that represents the perimeter of Karen's garden.
   b. Write a linear equation that describes the length of the garden in terms of the width.
   c. Graph both equations on the same coordinate plane.
   d. What are the length and width of Karen's garden?

20. Barry and Helen each own sailboats that are docked in different locations. Both decide to go sailing on Saturday morning and leave at the same time. Barry's sailing path can be described by the linear equation $y = 12x - 30$ and Helen's path can be described by the equation $y = 3x + 60$. At what point will they cross paths? Show all work necessary to justify your answer.

21. Javier is solving a system of equations that has values in the solution which are not integers. Why might solving the system by graphing not give him the most accurate answer? Explain your reasoning.

22. Missy says her system of linear equations has exactly two solutions, (2, 8) and (7, −5). Is this possible? Explain your reasoning.

REVIEW

Determine if the two lines in each system of equations are intersecting, parallel or the same line. State how many solutions there will be for each system. Use words and/or numbers to show how you determined your answer.

23. $y = -2x - 5$
$y = 2x + 1$

24. $4x - 8y = 16$
$y = \frac{1}{2}x - 2$

25. $y = 5(x + 3) - 1$
$y = 5x + 7$

26. $y = \frac{2}{3}x - 1$
$-2x + 3y = 3$

27. $y = \frac{1}{3}x + 3$
$x + 2y = 6$

28. $y = 3(2x + 1) - 5$
$y = 6(x - 1) + 4$

SOLVING SYSTEMS USING TABLES

LESSON 4.3

 Determine the solution to a system of equations using tables.

Input-output tables are a tool used in mathematics to display information. Systems of equations can be solved using tables. An input-output table must be created for each equation in the system. The tables can be compared to find an (x, y) pair that is the same in each table. This point represents the solution to the system of equations.

EXPLORE! LARRY'S LANDSCAPING

Larry's Landscaping offers two payment options for his employees. Option #1 offers $925 per month in salary plus $25 for every job completed. Option #2 offers $1,000 every month plus $10 for every job completed.

Step 1: Write an equation to represent the monthly salary, y, that could be earned for x jobs completed if an employee chooses Option #1.

Step 2: Write an equation to represent the monthly salary, y, that could be earned for x jobs completed if an employee chooses Option #2.

Step 3: Copy the two tables shown below on your own paper. Calculate the monthly salary for an employee under each plan for 0 through 10 jobs.

Option #1

Jobs Completed, x	Monthly Salary, y
0	
1	
2	

Option #2

Jobs Completed, x	Monthly Salary, y
0	
1	
2	

Step 4: The solution to this system of equations occurs when an employee earns the same amount of money for the same number of jobs. Use your table to determine when this happens. Write your answer in a complete sentence.

Step 5: Verify your answer by substituting the x- and y-values of your solution into the original equations in the system to see if the ordered pair makes each equation true.

Step 6: If an employee thinks he can complete 50 jobs in one month, which payment option should he choose? Explain your answer.

SOLVING SYSTEMS OF LINEAR EQUATIONS USING TABLES

1. Convert both linear equations in the system to slope-intercept form.
2. Create an input-output table for each equation. Use the same input values for each table.
3. Locate the point in each table where the same pair of input and output values occurs. This is the solution to the system of equations.
4. Verify that the ordered pair is the solution by substituting the x- and y-values into both equations in the system.

EXAMPLE 1

Solve the system of equations using input-output tables. Check the solution.
$$y = -3x + 13 \qquad 2x + y = 9$$

SOLUTION

Convert both equations to slope-intercept form:

$y = -3x + 13$
↓
$\boxed{y = -3x + 13}$

$2x + y = 9$
$-2x \qquad -2x$
↓
$\boxed{y = 9 - 2x}$

Create an input-output table for each equation using the same input values.

$y = -3x + 13$

x	y
0	13
1	10
2	7
3	4
4	1
5	-2

$y = 9 - 2x$

x	y
0	9
1	7
2	5
3	3
4	1
5	-1

> The solution is the ordered pair that occurs in both tables.

The solution to the system of equations is (4, 1).

☑ Verify the answer by substituting 4 for x and 1 for y in the original equations.

$y = -3x + 13$
$1 \stackrel{?}{=} -3(4) + 13$
$1 \stackrel{?}{=} -12 + 13$
$1 = 1$

$2x + y = 9$
$2(4) + 1 \stackrel{?}{=} 9$
$8 + 1 \stackrel{?}{=} 9$
$9 = 9$

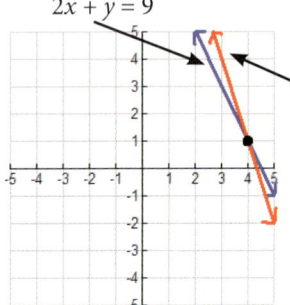

A solution to a system of equations that is solved using input-output tables can also be verified using a graph. In **Example 1**, the solution to the system of equations is (4, 1). This means the lines intersect at (4, 1).

EXAMPLE 2 Solve the system of equations using input-output tables. Check the solution.
$$y = \tfrac{1}{2}x + 4 \qquad y = -x + 1$$

SOLUTION Both equations are in slope-intercept form. Create an input-output table for each equation using the same input values.

$y = \tfrac{1}{2}x + 4$

x	y
0	4
1	$4\tfrac{1}{2}$
2	5
3	$5\tfrac{1}{2}$

y-values are getting larger.

$y = -x + 1$

x	y
0	1
1	0
2	−1
3	−2

y-values are getting smaller.

The *y*-values are going in opposite directions. The solution must have a negative *x*-value. Use input values that are negative to find the solution.

$y = \tfrac{1}{2}x + 4$

x	y
−1	$3\tfrac{1}{2}$
−2	3
−3	$2\tfrac{1}{2}$
−4	2

$y = -x + 1$

x	y
−1	2
−2	3
−3	4
−4	5

The solution to the system of equations is (−2, 3).

☑ Verify the answer by substituting −2 for *x* and 3 for *y* in the original equations.

$y = \tfrac{1}{2}x + 4$ $y = -x + 1$
$3 \stackrel{?}{=} \tfrac{1}{2}(-2) + 4$ $3 \stackrel{?}{=} -(-2) + 1$
$3 \stackrel{?}{=} -1 + 4$ $3 \stackrel{?}{=} 2 + 1$
$3 = 3$ $3 = 3$

EXERCISES

Solve each system of equations using the given input-output tables. Show all work necessary to justify your answer.

1.

$y = 3x - 1$

x	y
0	
1	
2	
3	

$y = -2x + 4$

x	y
0	
1	
2	
3	

2.

$y = x + 6$

x	y
−3	
−2	
−1	
0	

$y = \tfrac{1}{2}x + 5$

x	y
−3	
−2	
−1	
0	

3. Abe created input-output tables with input-values of 0, 1, 2, 3 and 4 to solve his system of equations. After looking at his output values, he realized he needed to try negative input values. What do you think he noticed about his output values?

Solve each system of equations using input-output tables. Show all work necessary to justify your answer.

4. $y = 5x - 6$
$y = -2x + 15$

5. $y = x + 2$
$y = 2x + 1$

6. $y = 3x + 4$
$y = 2x + 14$

7. $y = \frac{1}{2}x$
$y = 3x + 10$

8. $2x + y = 3$
$y - 3x = 23$

9. $-4x + 2y = 12$
$y = -5x + 6$

10. Two submarines were headed toward one another. One followed the path represented by the equation $y = 4x + 7$. The other submarine followed the path represented by the equation $y = 3x + 12$. Let x represent the number of minutes the submarines have been in motion and y represent the distance each is from the submarine base.
   a. Solve the system of equations using two input-output tables to determine when the submarines' paths will cross.
   b. Explain how you know your answer is correct.

11. Carlos put $100 in a savings account at the beginning of the year. At the end of each month, he added $15 to the account. Ana put $400 in her savings account at the beginning of the year. At the end of each month, she took $35 out of her account. Let x represent the number of months which have passed and y represent the amount in each savings account.
   a. Write an equation to represent the amount in Carlos' savings account.
   b. Write an equation to represent the amount in Ana's savings account.
   c. Copy and complete the input-output tables through 8 months.

Carlos' Savings Account Balance

Months, x	Total Savings, y
0	
1	
2	

Ana's Savings Account Balance

Months, x	Total Savings, y
0	
1	
2	

   d. When will Carlos and Ana have the same amount in their savings accounts? How much will they each have at this time?

12. Solve each system of equations using input-output tables. Verify each solution by graphing the two equations.
   a. $y = \frac{1}{2}x - 1$
   $y = -x + 5$

   b. $y = -2x - 3$
   $y = x - 9$

13. Joshua's profits, P, for his lawn-mowing business are represented by the equation $P = 16m - 52$ where m is the number of lawns he has mowed. Serj also runs a lawn-mowing business. His profits can be calculated using the equation $P = 14m - 40$. How many lawns do they have to mow to make the same amount of profit? Explain how you know your answer is correct.

REVIEW

Solve each equation. Show all work necessary to justify your answer.

14. $8x - 10 = -2x + 60$

15. $\frac{2}{3}x + 7 = \frac{4}{3}x + 6$

16. $4x + 2 = 5x + 7$

17. $-2x = 6x + 40$

18. $x + 3 = \frac{1}{2}x + 1$

19. $3.2x - 12 = 4.7x - 9$

Tic-Tac-Toe ~ Different Systems

Systems of equations can include equations that are not linear. In this activity, you will be finding the solutions to systems of equations containing a linear equation and a quadratic equation. Each system will have 2 solutions. You may use graphing or input-output tables to find the solutions.

For example: $y = x + 3$
$y = x^2 - 3$

$y = x + 3$

x	y
-3	0
-2	1
-1	2
0	3
1	4
2	5
3	6

$y = x^2 - 3$

x	y
-3	6
-2	1
-1	-2
0	-3
1	-2
2	1
3	6

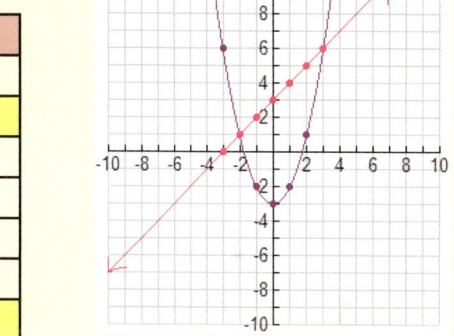

SOLUTIONS: $(-2, 1)$ and $(3, 6)$

Find the two solutions to each system of equations. Show all work.

1. $y = x^2$
$y = 2x + 3$

2. $y = -x^2 + 1$
$y = x - 5$

3. $y = 3x + 1$
$y = x^2 + 1$

Tic-Tac-Toe ~ Math Dictionary

Create a "Linear Equations" Dictionary. Locate all of the vocabulary words from the first four Blocks in this textbook. Alphabetize the list of words and design a dictionary. The dictionary should include each word, spelled correctly, along with the definition. If appropriate, a diagram or illustration can be included.

Lesson 4.3 ~ Solving Systems using Tables

SOLVING SYSTEMS BY SUBSTITUTION

LESSON 4.4

 Determine the solution to a system of equations using the substitution method.

Solving a system of linear equations by graphing or input-output tables is convenient when the ordered pair solution contains only small integers. This will not occur with every system of equations. This lesson shows another method for solving a system of linear equations called the **substitution method**.

SOLVING SYSTEMS OF LINEAR EQUATIONS BY SUBSTITUTION

1. Solve one of the linear equations for a variable (isolate x or y), if necessary.
2. Replace the variable in the second equation with the expression that you solved for in **Step 1**. Solve for the variable in your new equation.
3. Substitute your solution into the equation from **Step 1** to find the value of the other variable. State your full answer as an ordered pair (x, y).
4. Verify that the ordered pair is the solution by substituting the x- and y-values into both equations in the system or by graphing the system to confirm that the point of intersection matches your solution.

EXAMPLE 1 Use the substitution method to solve the system of linear equations.
$$y = -2x + 5$$
$$4x + 3y = 9$$

SOLUTION The first equation has an isolated variable. Since $y = -2x + 5$, substitute $-2x + 5$ for y in the second equation and solve for x.

$$4x + 3y = 9$$
$$4x + 3(-2x + 5) = 9$$
$$4x + -6x + 15 = 9$$
$$-2x + 15 = 9$$
$$\, -15 \; -15$$
$$\frac{-2x}{-2} = \frac{-6}{-2}$$
$$x = 3$$

Always put the expression in parentheses because you will often have to use the Distributive Property.

Substitute 3 for x in the first equation.

$$y = -2(3) + 5$$
$$y = -6 + 5$$
$$y = -1$$

Verify that $(3, -1)$ makes both equations true.

$$y = -2x + 5 \qquad\qquad 4x + 3y = 9$$
$$-1 \stackrel{?}{=} -2(3) + 5 \qquad\qquad 4(3) + 3(-1) \stackrel{?}{=} 9$$
$$-1 \stackrel{?}{=} -6 + 5 \qquad\qquad 12 + -3 \stackrel{?}{=} 9$$
$$-1 = -1 \qquad\qquad 9 = 9$$

Solution: $(3, -1)$

You can also graph the two linear equations to verify that your solution matches the point of intersection. Looking at this graph you can see why substitution was a better method than graphing. It is hard to determine the exact point of intersection on the graph.

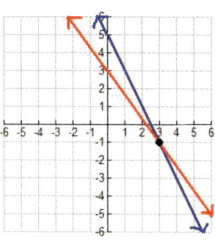

EXPLORE!

A TRIP ON I-70

Zach lives in Salina, Kansas. Gina lives 45 miles east in Junction City, Kansas. Gina and Zach both leave their houses at the same time, heading east on I-70. Zach drives 65 miles per hour and Gina drives 55 miles per hour. Zach wants to determine how long it will take before he catches up with Gina.

Step 1: Copy the equations listed below and identify which equation corresponds to Zach and which equation corresponds to Gina. The y-variable represents the distance from Zach's house. Describe what the x-variable represents in this situation.

$$y = 55x + 45$$
$$y = 65x$$

Step 2: Solve this system of equations using the substitution method.

Step 3: How many hours will it take before Zach catches Gina? How far will Zach have driven at this point?

Step 4: Explain how you know your answer is correct.

EXERCISES

Identify the equation that has an isolated variable. State which variable is isolated.

1. Equation #1 $\quad 3x - 5y = 10$
Equation #2 $\quad x = 4 - 4y$

2. Equation #1 $\quad y = \frac{1}{2}x - 4$
Equation #2 $\quad x - 5y = 5$

3. Equation #1 $\quad -x + 3y = 10$
Equation #2 $\quad y = 4 - 2x$

4. Equation #1 $\quad 12 - 6y = x$
Equation #2 $\quad 2x + y = 6$

5. Nathan wants to know why he cannot always just use graphing to solve a system of linear equations. How would you answer his question?

6. Explain two different methods that can be used to verify an answer when solving a system of equations using substitution.

7. Isolate the y-variable in the equation $2x + y = 7$.

8. Isolate the x-variable in the equation $3x = 6 + 9y$.

Solve each system of equations using the substitution method. Show all work necessary to justify your answer.

9. $x = y - 3$
$5x + 3y = 1$

10. $3x - y = 7$
$y = 2x - 4$

11. $y = 10 - 2x$
$3x - 2y = 22$

12. $x = 12 + 3y$
$2x + 5y = -20$

13. $y = 15 + x$
$2x + 5y = 26$

14. $5y + 7x = 2$
$x = 2 + y$

15. $y = -5 + x$
$-2x + y = -4$

16. $2x + 3y = 17$
$2x + y = 3$

17. $x - 7y = 4$
$3x + y = -10$

18. Hank's Ice Cream Shop sells single and double scoop cones. The single-scoop cones cost $2.00 and the double-scoop cones cost $2.50. In one day he sold 230 cones for a total of $498 in sales.
   a. Explain why the equations $x + y = 230$ and $2x + 2.5y = 498$ represent this situation.
   b. What does x represent in the equations in **part a**? What does y represent?
   c. Isolate one variable in an equation. Solve the system of equations using the substitution method.
   d. What is the real-world meaning of the solution to the system?

19. Emma picked two numbers, x and y. She told her teacher that the sum of the two numbers was 46 and the difference of the two numbers was 12.
   a. Write two different linear equations that model what Emma told her teacher.
   b. Isolate one variable in an equation. Solve the system of linear equations using the substitution method. What were the two numbers Emma picked?

20. Aiden solved a system of equations using substitution. He stated that his solution was $x = 7$. Francis said something seemed wrong about the solution because it was not a point on a graph but just an x-value. Do you think Aiden's solution is complete? If not, what else does he need to do?

21. Solve the system of equations shown below using three methods (graphing, input-output tables and substitution).
$$y = 2x + 3$$
$$y = 15 - x$$

22. The Flying W Ranch raises only cows and horses. There are a total of 340 animals on the ranch. The owner prefers horses over cows so he has 52 more horses than cows.
   a. Write two different linear equations to model this situation.
   b. Solve the system of linear equations using the substitution method. How many horses live at the Flying W Ranch?

23. Tad and Timothy went to the paint store together. Tad bought 6 cans of paint and 1 paint brush for $67. Timothy bought 4 cans of the same paint and 3 of the same type of paint brushes. Timothy's total cost was $54.

    a. Write a linear equation that represents Tad's purchase and another to represent Timothy's purchase. Let x represent the cost of a can of paint and y represent the cost of a paint brush.

    b. Solve the system of linear equations using the substitution method.

    c. What was the cost for a can of paint? The cost of a paint brush?

REVIEW

Determine if the two lines in each system of equations are intersecting, parallel or the same line. State how many solutions there will be for each system.

24. $y = \frac{4}{5}x + 3$
    $y = \frac{4}{5}x - 3$

25. $y = \frac{1}{2}x + 5$
    $y = -\frac{1}{2}x + 5$

26. $y = 2(x + 1)$
    $y = 2x + 1$

27. $4x + 2y = 30$
    $y = \frac{4}{3}x - 5$

28. $-x + 6y = 6$
    $y = \frac{1}{6}x + 1$

29. $y = 6(x + 1) - 4$
    $6x - y = 2$

Tic-Tac-Toe ~ Letter to Fifth Graders

Write a letter to a class of fifth grade students explaining why it is important to learn math. Support your reasons with research. Give some examples of real-world situations in which they will encounter math. Include any advice you believe would help them be successful in mathematics through the middle school years. Turn in one copy of the letter to your teacher and give another copy of the letter to a fifth grade teacher in your district.

Tic-Tac-Toe ~ Polygons

Polygons are enclosed figures whose sides are made up of line segments. Create a polygon (triangle, quadrilateral, pentagon, hexagon, etc.) by graphing linear equations that enclose the polygon. Write the equations for each line. List the vertices (or points of intersection). Color in the polygon. Repeat the process on another sheet of graph paper, creating a different polygon.

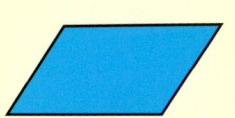 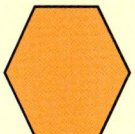    

SOLVING SYSTEMS USING ELIMINATION

LESSON 4.5

 Determine the solution to a system of equations using the elimination method.

So far in this block, you have used three methods to solve systems of linear equations: graphing, tables and the substitution method. There are times when both equations in a system may be in standard form ($Ax + By = C$). When this occurs, the easiest method to use to solve the system will be the **elimination method**. The elimination method involves combining the two equations in a way that will "eliminate" one of the variables so that you can solve for the remaining variable.

SOLVING SYSTEMS OF LINEAR EQUATIONS BY ELIMINATION

1. Arrange the equations so the common variables are lined up vertically in columns and the constants are alone on one side of the equals sign.
2. Multiply one or both equations so that one of the variables (x or y) have coefficients that are opposites, if necessary.
3. Add the columns together. One variable should cancel out by adding to zero. Solve for the remaining variable.
4. Substitute your solution into either of the original equations and solve for the other variable.
5. Verify that the ordered pair is the solution by substituting the x- and y-values into both equations in the system or by graphing.

EXAMPLE 1 Use the elimination method to solve the system of linear equations.
$$3x - 2y = 1$$
$$2x + 2y = 4$$

SOLUTION

The equations are arranged properly with the variables lined up vertically.
$$3x - 2y = 1$$
$$2x + 2y = 4$$

The equations have y-variables that have coefficients of 2 and −2. These are opposites so y is the variable that will be eliminated.

Add the columns together. The y-variable is eliminated. Solve for the remaining variable, x.

$$3x - 2y = 1$$
$$\underline{2x + 2y = 4}$$
$$\frac{5x}{5} = \frac{5}{5}$$
$$x = 1$$

Substitute the value of x into one of the original equations.

$$2(1) + 2y = 4$$
$$2 + 2y = 4$$
$$\underline{-2 \quad -2}$$
$$\frac{2y}{2} = \frac{2}{2}$$
$$y = 1$$

Solution (1, 1)

EXAMPLE 1
SOLUTION
(CONTINUED)

☑ Verify that (1, 1) satisfies both equations.

$$3x - 2y = 1 \qquad\qquad 2x + 2y = 4$$
$$3(1) - 2(1) \stackrel{?}{=} 1 \qquad\qquad 2(1) + 2(1) \stackrel{?}{=} 4$$
$$3 - 2 \stackrel{?}{=} 1 \qquad\qquad 2 + 2 \stackrel{?}{=} 4$$
$$1 = 1 \qquad\qquad 4 = 4$$

EXAMPLE 2

Use the elimination method to solve the system of linear equations.

$$3x + y = 7$$
$$2x + 5y = 22$$

SOLUTION

The equations are arranged properly with the variables lined up vertically.

$$3x + y = 7$$
$$2x + 5y = 22$$

Neither the x- or y-terms have coefficients that are opposites of each other. In order for the x-coefficients to be opposites, both equations would have to be multiplied by constants (the first by 2 and the second by −3). The y-coefficients are easier to make opposites since only one equation needs to be multiplied by a constant. Multiply one equation through by a constant that will create opposites.

$$3x + y = 7 \quad\rightarrow\quad -5(3x + y = 7) \quad\rightarrow\quad -15x + -5y = -35$$
$$2x + 5y = 22 \quad\rightarrow\quad\quad\quad\quad\quad\quad\quad\rightarrow\quad 2x + 5y = 22$$

Choose −5 so the coefficients of the y-variables are opposites.

Add the columns together. The y-variable is eliminated. Solve for the remaining variable.

$$-15x + -5y = -35$$
$$\underline{2x + 5y = 22}$$
$$\frac{-13x}{-13} = \frac{-13}{-13}$$
$$x = 1$$

Substitute the value of x into one of the original equations.

$$3(1) + y = 7$$
$$3 + y = 7$$
$$\underline{-3 \quad\quad -3}$$
$$y = 4$$

Solution (1, 4)

Verify that (1, 4) satisfies both equations.

$$3x + y = 7 \qquad\qquad 2x + 5y = 22$$
$$3(1) + 4 \stackrel{?}{=} 7 \qquad\qquad 2(1) + 5(4) \stackrel{?}{=} 22$$
$$3 + 4 \stackrel{?}{=} 7 \qquad\qquad 2 + 20 \stackrel{?}{=} 22$$
$$7 = 7 \qquad\qquad 22 = 22$$

There are four types of steps to follow when trying to get a system of equations ready for "eliminating" a variable.

ZERO-STEP: There is one variable in each equation whose coefficients are opposites.

$$x - 2y = 2$$
$$x + 2y = 10$$

When the equations are added together, the y terms will cancel.

ONE-STEP: Neither set of variables have opposite coefficients, but one term is a multiple of its corresponding term.

The x term in the bottom equation is a multiple of the x term in the first equation. Both are positive so multiply the top equation through by –4.

$$x + 3y = -12 \rightarrow -4(x + 3y = -12) \rightarrow -4x - 12y = 48$$
$$4x - 5y = 37 \rightarrow \rightarrow 4x - 5y = 37$$

SPECIAL TYPE OF ONE-STEP: There is one variable in each equation whose coefficients are equal, not opposites.

The x terms match but they are both positive. Multiply one equation by –1 and distribute.

$$3x + 3y = -3 \rightarrow -1(3x + 3y = -3) \rightarrow -3x - 3y = 3$$
$$3x + 2y = -1 \rightarrow \rightarrow 3x + 2y = -1$$

TWO-STEP: Neither set of variables have opposite coefficients. In order to get opposite coefficients on one variable, both equations must be multiplied by different numbers.

Either variable could be chosen to eliminate in this system (y was chosen because one was positive and one was negative). Multiply each equation through in order to reach the least common multiple.

$$2x - 5y = 7 \rightarrow 2(2x - 5y = 7) \rightarrow 4x - 10y = 14$$
$$3x + 2y = 20 \rightarrow 5(3x + 2y = 20) \rightarrow 15x + 10y = 100$$

EXERCISES

Show all work necessary to transform each system of linear equations into a system that is ready for columns to be added together to eliminate a variable. Rewrite the system if it is already set up for elimination.

1. $x + 4y = 23$
$-x + y = 2$

2. $2x + 3y = 13$
$x - 2y = 2$

3. $5x + 2y = 6$
$9x + 2y = 22$

4. $2x + y = 4$
$5x + 4y = 7$

5. $3x + 2y = 8$
$2x - 3y = -12$

6. $2x + 7y = -3$
$x + y = -4$

7. Alexandra wants to eliminate the x-variable in the system below. She multiplies the first equation by 3 and the second equation by 5. She adds the two equations together and the x-variables are still there. What did she do wrong?

$$5x + 3y = 10 15x + 9y = 30$$
$$3x + 5y = 14 15x + 25y = 70 30x + 34y = 100$$

Solve each system of equations using the elimination method. Show all work necessary to justify your answer.

8. $x + y = 11$
$x - y = -3$

9. $3x + 2y = 0$
$-3x + y = 9$

10. $5x + 2y = -5$
$-x + 3y = 1$

11. $6x - 2y = 36$
$3x - 2y = 21$

12. $3x + y = -11$
$4x - 3y = 7$

13. $5x + 2y = -1$
$x - 2y = 1$

14. $7x - 4y = 26$
$5x + 4y = 46$

15. $12x + 3y = 12$
$8x - 2y = 4$

16. $3x + 4y = -25$
$2x - 3y = 6$

17. Two children's blocks are chosen. Three times the value of one block plus the value of the second block is 29. The value of the first block plus twice the value of the second block is 18.
   a. Write two equations using the information given about the two chosen blocks to create a system of linear equations.
   b. Solve the system of equations using the elimination method. What is the value of each chosen block?

18. Irina sells two types of candy bars for a fundraiser. One type costs $1 and the other costs $2. At the end of the fundraiser, she has sold 44 candy bars for a total of $68. She wants to determine the number of each type of candy bar she has sold. Let x represent the number of $1 candy bars she has sold and y represent the number of $2 candy bars she has sold.
   a. Explain why the equations $x + y = 44$ and $x + 2y = 68$ represent this situation.
   b. Solve the system of equations using the elimination method.
   c. How many $1 candy bars did she sell? How many $2 candy bars did she sell?

REVIEW

Solve each system of linear equations using the graphing method.

19. $y = \frac{4}{5}x - 2$
$y = -\frac{2}{5}x + 4$

20. $y = x - 1$
$y = -3x - 5$

21. $y = -\frac{1}{2}x + 2$
$y = \frac{1}{2}x - 2$

Solve each system of linear equations using the substitution method.

22. $4x + 3y = 31$
$x = 2y + 5$

23. $x = 15 + 12y$
$2x + 3y = 3$

24. $2x + 5y = -1$
$x + y = 4$

Tic-Tac-Toe ~ Solution Given

Below are solutions to systems of two linear equations. Create a system of linear equations that has each solution. You may not use any horizontal or vertical lines. Prove that your system of equations has the solution by solving the system using graphing, tables, substitution or elimination.

1. (3, 2)
2. (0, 5)
3. (−1, 4)
4. (2, −6)
5. (−3, −3)
6. (8, 0)

CHOOSING THE BEST METHOD

LESSON 4.6

 Choose the best method for solving a given system of equations.

You have learned four different ways to solve a system of linear equations. All four methods work on any problem, but each problem is usually set up in a way that makes one method of solving easier than the other methods.

A quick review of the four methods you have learned to solve systems of linear equations:

GRAPHING	Graph both equations in slope-intercept form. Identify their point of intersection.
TABLES	Create an input-output table for each equation using the same input values. Locate the point in each table where the same pair of input and output values occur.
SUBSTITUTION	Solve for a variable in one equation and substitute that expression into the other equation. Solve that equation for one variable. Substitute the solution into one of the original equations to solve for the second variable.
ELIMINATION	Create opposite coefficients on one variable in the two equations. Add the equations together to eliminate one variable. Solve for the remaining variable. Substitute the solution into one of the original equations to solve for the second variable.

EXPLORE! — WHAT'S EASIEST?

Nate and Tabi were given five systems of equations to solve. Their teacher told them each system could be solved using any of the four methods above. For each one, however, there is one method that would be the easiest to use.

System #1
$3x + y = 12$
$-3x + y = 30$

System #2
$y = \frac{1}{2}x - 5$
$y = -x + 1$

System #3
$y = 2x + 5$
$y = 4x + 5$

System #4
$y = 2x + 1$
$4x + 5y = -9$

System #5
$2x - 3y = 12$
$x = 2y + 8$

Step 1: Tabi likes elimination the best so she decides that she will solve them all with elimination. Her teacher said one of the systems is set up for elimination. Which system do you think the teacher referred to? Explain your reasoning.

Step 2: Nate believes substitution is always the easiest method to use, no matter how the system is set up. The teacher told Nate there are two systems set up in a way that will make substitution the easiest method for solving. Which two systems do you think the teacher was talking about and why?

Step 3: Two systems are left. Which one would you solve by graphing? Which one would you solve using input-output tables? Explain your reasoning.

Step 4: Choose one of the systems above and find the solution.

Choosing A Method To Solve a System of Linear Equations

Graphing: If both equations are in slope-intercept form, graphing is an appropriate method to use.

Tables: If both equations are in slope-intercept form and the slope is an integer, input-output tables may be a good method.

Substitution: If a variable is isolated in one equation, substitution will most likely be the best method.

Elimination: If the two equations in the system are in standard form and the variables are lined up in columns, elimination may be the easiest method.

EXAMPLE 1

Choose the best method to solve each system of linear equations. Explain your reasoning.

a. $x - 2y = 10$
$3x + 2y = 6$

b. $y = 2x - 3$
$4x - 5y = -1$

c. $y = \frac{1}{3}x$
$y = 2x + 7$

d. $y = 5x - 9$
$y = 2x - 3$

SOLUTIONS

a. The best method for solving this system of linear equations would be **ELIMINATION**. This method would be easiest because the variables are already lined up in columns and the y–variable is already set to be eliminated when the two equations are added together.

b. The best method for this system is **SUBSTITUTION**. The y-variable in the first equation is isolated on one side of the equals sign which provides an expression to substitute into the other equation and solve.

c. Both equations are solved for y which allows this to be easily solved by **GRAPHING**. Remember that when the graphing method is used, it is important to verify the solution by substituting the x-and y-values back into both equations.

d. Both equations are in slope-intercept form. The slope in each equation is an integer. The best method for solving this system of equations may be **TABLES**.

Lesson 4.6 ~ Choosing the Best Method

EXERCISES

Choose the best method to solve each system of linear equations. Explain your reasoning.

1. $3x + y = 13$
   $-3x - 4y = -7$

2. $x = 4y$
   $3x + 2y = 11$

3. $y = 3x - 7$
   $y = -\frac{3}{2}x + 2$

4. Lucy chose to use the graphing method to solve the system of equations shown below. Her friend, Dave, argues that graphing is not the best method. He says he would use tables. Who do you agree with and why?

   $y = 40x + 290$
   $y = 500 + 25x$

State the best method to solve each system of linear equations. Solve the system.

5. $3x - y = 14$
   $x + y = 2$

6. $x = y + 3$
   $2x + 3y = 1$

7. $-4x + 2y = -6$
   $2x - 5y = -9$

8. $y = 2x - 4$
   $y = x + 1$

9. $y = \frac{1}{2}x + 3$
   $3x - 4y = -10$

10. $y = \frac{1}{2}x + 3$
    $y = -3x - 4$

11. $3x + 2y = -18$
    $-2x + 5y = -26$

12. $y = -\frac{3}{2}x + 5$
    $y = 2x - 9$

13. $y = 5x - 2.5$
    $4x + 2y = 9$

REVIEW

14. Listed below are six linear equations related to the line $y = \frac{2}{3}x - 4$. Two of the lines are parallel to this line. Two of the lines are the exact same line. The other two intersect the line. Determine which lines fit in each category. Use words and/or numbers to show how you determined your answer.

Line A
$-3x + 2y = 6$

INTERSECTING?

Line C
$y = \frac{2}{3}(x + 3) - 6$

Line D
$4x - 6y = 24$

Line B
$y = \frac{2}{3}x + 4$

Line F
$y = -\frac{2}{3}x - 4$

THE SAME LINE?

Line E
$y = \frac{2}{3}(x + 6) - 5$

PARALLEL?

APPLICATIONS OF SYSTEMS OF EQUATIONS

LESSON 4.7

 Set up and solve systems of equations from word problems.

Systems of linear equations are used to solve problems in all types of real-world situations. In this lesson, you will see systems of linear equations used to solve problems involving cell phone plans, job options and shipping costs. You will be given details about a problem that will provide you with enough information to write two linear equations. You must first determine what each variable represents. Once the two linear equations are developed, you must determine which method you want to use to solve the system. Always remember to check your answer by referring back to the original problem to see if your solution is correct.

EXPLORE!   AT THE MOVIES

The Rodriguez family and the Jacobson family go to the movies together. The Rodriguez family bought 3 adult tickets and 2 youth tickets for a total of $29.00. The Jacobson family bought 2 adult tickets and 5 youth tickets for a total of $31.25. Let x represent the cost of an adult ticket and y represent the cost of a youth ticket.

Step 1: Write an equation to represent the Rodriguez family's movie ticket purchase.

Step 2: Write an equation to represent the Jacobson family's movie ticket purchase.

Step 3: Choose the best method for solving this system of equations. Why did you choose that method?

Step 4: Solve your system of linear equations.

Step 5: How much did a youth's ticket cost at this movie theater? What was the cost of an adult ticket?

Step 6: Check your answer by determining if your ticket prices give the same totals that were charged to the Jacobson and Rodriguez family.

Step 7: The Chang family also went to see the same movie as the other two families. The Changs bought one adult ticket and 3 youth's tickets. What was the total cost for the Chang family to go to the movies? Use words and/or numbers to show how you determined your answer.

Lesson 4.7 ~ Applications of Systems of Equations

EXAMPLE 1

Nai is trying to decide between two different cell phone plans. Plan A charges a flat fee of $22 per month plus $0.10 per minute of phone usage. Plan B charges $0.18 per minute with no flat fee.

a. How many minutes would Nai have to use each month for the cell phone plans to cost the same amount? How much would it cost?

b. If Nai figures he will talk 400 minutes on the phone each month, which plan should he choose?

SOLUTIONS

a. Let x represent the minutes talked and y represent the total monthly cost.

Write a system of two linear equations to model this situation.

Plan A: $y = 22 + 0.10x$ **Plan B:** $y = 0.18x$

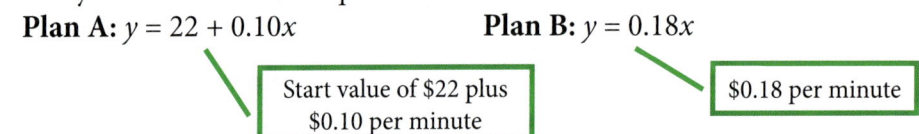

Start value of $22 plus $0.10 per minute

$0.18 per minute

Choose a method and solve the system of equations. Graphing or substitution would work. SUBSTITUTION will work best because it will provide an accurate answer since it is not known if the solution will have integer values.

Solve the system.

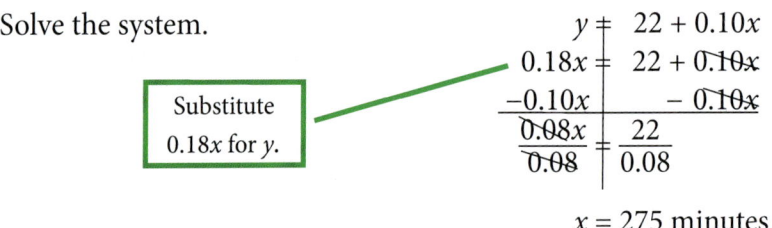

Substitute $0.18x$ for y.

$$y = 22 + 0.10x$$
$$0.18x = 22 + 0.10x$$
$$-0.10x \quad\quad -0.10x$$
$$\frac{0.08x}{0.08} = \frac{22}{0.08}$$

$x = 275$ minutes

Substitute the x-value into one of the original equations to determine the total cost when the plans cost the same.
$$y = 0.18(275) = \$49.50$$

The plans would cost the same amount, $49.50, after 275 minutes.

b. Nai plans to talk 400 minutes each month. To determine which plan is best, substitute 400 for x in each equation to see which plan will be less expensive.

Plan A: $y = 22 + 0.10(400) = \$62$
Plan B: $y = 0.18(400) = \$72$

Nai should choose **Plan A** if he plans to use his phone 400 minutes each month.

EXAMPLE 2

Omar has two possible sales job options. Job Option #1 has a monthly salary of $1,200 plus 4% of his total sales. Job Option #2 has a monthly salary of $1,500 plus 2% of his total sales.

a. How much would Omar have to sell to earn the same amount in one month at each job?
b. Omar thinks he can sell an average of $8,000 worth of merchandise in one month. Which job should he take?

SOLUTIONS

a. Let x represent the amount of Omar's sales in one month and y represent the total monthly salary.

Write a system of two linear equations to model this situation.
Job Option #1: $y = 1200 + 0.04x$ ──────── 4% = 0.04
Job Option #2: $y = 1500 + 0.02x$ ──────── 2% = 0.02

Choose a method and solve the system of equations. Graphing or substitution would work. SUBSTITUTION will work best because the y-intercepts are quite large and would be difficult to graph accurately.

Solve the system:

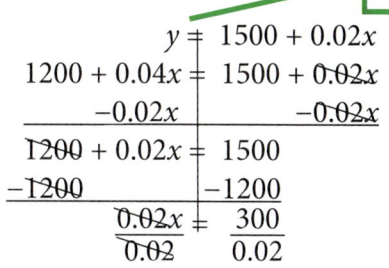

Substitute $1200 + 0.04x$ for y

$$y = 1500 + 0.02x$$
$$1200 + 0.04x = 1500 + 0.02x$$
$$-0.02x -0.02x$$
$$1200 + 0.02x = 1500$$
$$-1200 -1200$$
$$\frac{0.02x}{0.02} = \frac{300}{0.02}$$

$$x = \$15{,}000$$

Substitute the x-value into one of the original equations to determine the total salary when the jobs would pay the same.
$$y = 1200 + 0.04(15000) = \$1{,}800$$

Omar will earn the same monthly income ($1,800) at **Job Option #1 or #2** if he sells $15,000 worth of merchandise.

b. Use the original equations to determine which job will have the highest pay if he sells $8,000 worth of merchandise. Remember that his sales amount is substituted for x.

Job Option #1: $y = 1200 + 0.04(8000) = \$1{,}520$
Job Option #2: $y = 1500 + 0.02(8000) = \$1{,}660$

If Omar sells $8,000 worth of merchandise per month, he should take **Job Option #2.**

Lesson 4.7 ~ Applications of Systems of Equations

EXAMPLE 3

Sunshine Flower Company (SFC) ships boxes of tulip bulbs. The bulbs are always shipped in boxes that are the exact same size and weight. The billing statements are mailed in a separate envelope. On Monday, SFC shipped 210 boxes of bulbs and 140 billing statements. The total shipping bill for the day was $702.10. On Tuesday, SFC shipped 70 boxes of bulbs and 80 billing statements. The total shipping bill for Tuesday was $243.70. Determine the individual cost for mailing a box of bulbs and the cost for mailing a billing statement.

SOLUTION

Let x represent the cost of shipping a box of bulbs and y represent the cost of mailing a billing statement.

Write a system of two linear equations to model this situation.
Monday: $210x + 140y = 702.10$
Tuesday: $70x + 80y = 243.70$

Choose a method and solve the system of equations. This system of equations is set up for using the ELIMINATION method because the x- and y-variables are lined up in columns on one side of the equals sign and the constants are on the other side. Use multiplication in order to get opposite amounts of one variable.

Solve the system by first getting one variable with opposite coefficients.

$210x + 140y = 702.10$ → → $210x + 140y = 702.10$
$70x + 80y = 243.70$ → $-3(70x + 80y = 243.70)$ → $-210x - 240y = -731.10$

Add the columns together and solve for y.

$$\begin{array}{r} 210x + 140y = 702.10 \\ -210x - 240y = -731.10 \\ \hline \dfrac{-100y}{-100} = \dfrac{-29}{-100} \end{array}$$

$$y = \$0.29$$

Substitute the y-value into one of the original equations to solve for x:

$$70x + 80(0.29) = 243.70$$
$$70x + 23.20 = 243.70$$
$$-23.20 \quad -23.20$$
$$\dfrac{70x}{70} = \dfrac{220.50}{70}$$
$$x = \$3.15$$

Each box of bulbs costs $3.15 to ship. Each billing statement costs $0.29 to mail.

EXERCISES

Develop a system of equations for each problem. Describe what each variable represents.

1. Manny begins the summer with $200 in his savings account. Each week he adds $85 to his account. Susan begins the summer with $95 in a savings account and adds $100 each week. When will Manny and Susan have the same amount of money in their accounts?

2. Travis and Beth went to the corner mini-mart. Travis bought 6 candy bars and 2 sodas for $4.88. Beth bought 2 candy bars and 3 sodas for $3.47. All candy bars cost the same and all sodas are the same price. Determine the price of a candy bar and the price of a soda.

3. Otis wants to build a fence around his rectangular garden. The perimeter of the garden is 184 feet. The width is two times the length. Find the length and width of his garden. Use the formula for the perimeter of a rectangle as one of your equations.

Solve each problem using a system of equations. Define the variables and state the solution. Explain how you know your answer is correct.

4. Two taxi companies have different pricing systems. Company A charges a flat fee of $8 plus $0.10 per mile driven. Company B does not charge a flat fee, but charges $0.60 per mile driven. At what distance do both companies charge the same amount?

5. The set-up cost for a machine that attaches snaps on clothing is $1,100. After set-up, it costs $0.12 for each snap to be attached. A newer machine has come out that has a set-up cost of $1,520. With the new machine, it only costs $0.09 for each snap to be attached. How many snaps would the company have to attach to make the purchase of the newer machine worthwhile?

6. Two teachers, Mrs. Wright and Mr. Kinder, decide to buy calculators and protractors for their classrooms. Mrs. Wright buys 40 calculators and 30 protractors for $485. Mr. Kinder buys 20 of each for $252. What are the individual costs of the calculators and protractors that were purchased?

7. Two families had a garage sale together. The entire garage sale brought in $1,640. One family made $182 more than twice as much as the second family. How much did each family make at the garage sale?

8. On Friday afternoon, 560 people went to the local theater for the matinee. Youth tickets cost $5.75 and adult tickets cost $8.50. If the theater's sales receipts total $3,907.50, how many youth tickets were bought on Friday afternoon?

9. Jeremiah bought 3 gallons of ice cream and 4 containers of strawberries for $19.50. Gary bought 5 gallons of ice cream and 2 containers of strawberries for $22.00. What is the cost of one gallon of ice cream alone? What is the cost of one container of strawberries?

10. Two types of stereos were on sale at a local car stereo dealer. The J-Series model sold for $118. The K-Series model sold for $92. During the sale, 32 stereos were sold. The receipts for these stereos totaled $3,230. How many of each type of stereo did the local dealer sell during this sale?

11. Nancy and Pedro both drove to Nashville. Nancy started 72 miles closer to Nashville than Pedro did. Her average speed was 50 miles per hour. Pedro left at the same time Nancy left. He averaged 62 miles per hour. How long will it take before Pedro catches up with Nancy?

REVIEW

Graph each linear inequality.

12. $y > 2x - 5$

13. $y \leq \frac{3}{4}x + 1$

14. $y \geq -3x + 4$

15. $y < -\frac{6}{5}x + 7$

16. $y \geq -3$

17. $y < -2 + \frac{3}{2}x$

18. Write the inequality for the graph below.

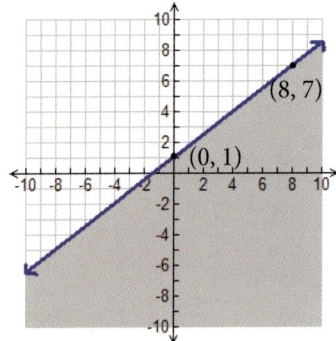

19. Write the inequality for the graph below.

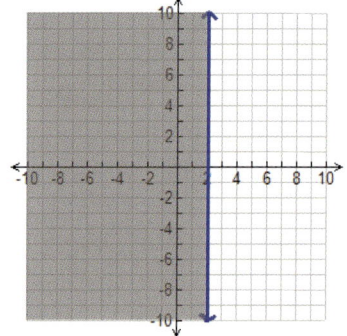

TIC-TAC-TOE ~ FRACTION COEFFICIENTS

Each system of linear equations below has at least one fraction coefficient.

Solve each system of linear equations using either substitution or elimination. Show all work necessary to justify each answer.

1. $y = \frac{1}{2}x + 4$
$2x + 4y = 40$

2. $\frac{1}{3}x + 2y = 3$
$x + 3y = 3$

3. $x = 4y - 2$
$\frac{1}{2}x + 5y = 6$

4. $\frac{1}{3}x + \frac{2}{3}y = 0$
$2x + \frac{1}{3}y = -11$

SYSTEMS OF LINEAR INEQUALITIES

LESSON 4.8

 Solve a system of linear inequalities by graphing.

In **Lesson 3.6**, you graphed linear inequalities in two variables by locating the boundary line and determining which side of the boundary line should be shaded. Just as two linear equations form a system of linear equations, two linear inequalities form a system of linear inequalities. In this lesson, you will graph two linear inequalities to find the region of ordered pairs that makes both inequalities true.

EXAMPLE 1 **Graph the solution to the system of linear inequalities:**
$$y \le \tfrac{1}{3}x - 2$$
$$y \ge -4x + 7$$

SOLUTION

Graph the boundary line by graphing $y = \tfrac{1}{3}x - 2$. Use a solid line to connect the points since the inequality symbol is $\le$.

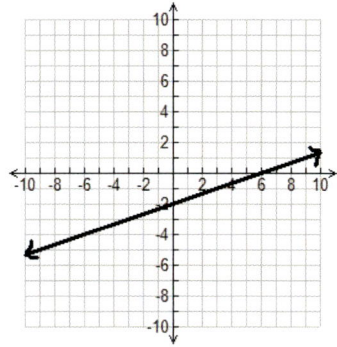

Choose (3, 6) as a test point. Substitute the coordinates into the original inequality to see if the statement is true or false.

$$6 \overset{?}{\le} \tfrac{1}{3}(3) - 2$$

Six is NOT less than or equal to −1 so the statement is false.

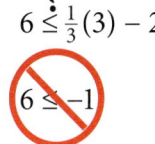

Shade the entire side (below the boundary line) that does not contains the test point (3, 6).

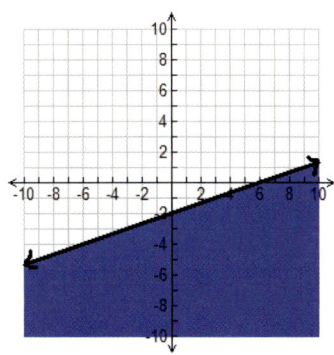

Continued on next page

EXAMPLE 1
SOLUTION
(CONTINUED)

Graph the boundary line of the second inequality by graphing $y = -4x + 7$. Use a solid line to connect the points since the inequality symbol is $\geq$.

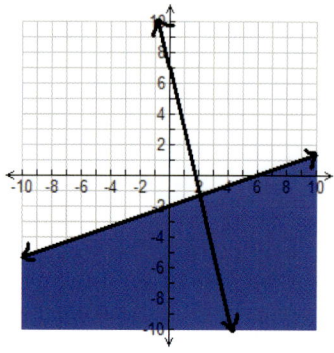

Choose (3, 6) as a test point. Substitute the coordinates into the original inequality to see if the statement is true or false.

$6 \overset{?}{\geq} -4(3) + 7$

Six is greater than -5 so the statement is true.

$6 \geq -5$

Shade the entire side that contains the test point (3, 6).

The solution to the system of inequalities is where the shaded regions overlap. The ordered pairs in this region make true statements in both inequalities.

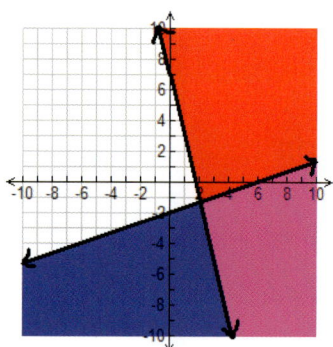

GRAPHING SYSTEMS OF LINEAR INEQUALITIES IN TWO-VARIABLES

1. Graph each linear inequality in the system on a single coordinate plane.
2. The solution to the system of inequalities is where the shaded regions overlap.

EXAMPLE 2

Is the point $(-4, -2)$ a solution to the system of inequalities formed by $y > x + 1$ and $y < 2x + 6$?

SOLUTION

Test the point $(-4, -2)$ in $y > x + 1$.

$y > x + 1$
$-2 > -4 + 1$
$-2 > -3$ **TRUE**

Test the point $(-4, -2)$ in $y < 2x + 6$.

$y < 2x + 6$
$-2 < 2(-4) + 6$
$-2 < -2$ **FALSE**

Since the point does not make both equations true, it is not a solution to the system of inequalities.

EXAMPLE 3 Graph the solution to the system of linear inequalities:
$y > -x + 5$
$y > 2$

SOLUTION Graph both inequalities on the same coordinate plane. Since the inequality symbols are both >, the boundary lines are dashed.

The solution set is the where the shaded regions overlap.

The points that fall on the dashed lines are not part of the solution set.

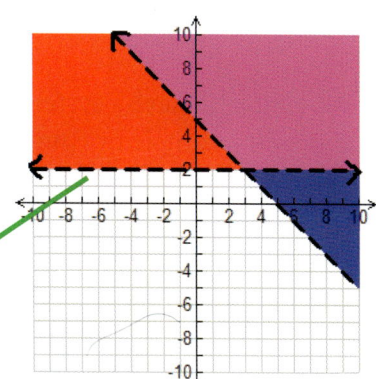

EXERCISES

1. On a graph representing a system of linear inequalities, how can you determine which region represents the solution to the system?

2. How are the solutions of systems of linear inequalities different from the solutions of systems of equations?

Graph the solution to each system of linear inequalities.

3. $y > -3x + 4$
$y < \frac{1}{3}x - 6$

4. $y \le x + 1$
$y > 5 - 2x$

5. $y \le \frac{2}{3}x$
$y \ge \frac{2}{3}x - 4$

6. $y > 2$
$y > -2$

7. $y \ge \frac{1}{4}x - 5$
$y < -\frac{5}{4}x + 1$

8. $y < 3 - 4x$
$y > 2x - 4$

9. $3x + 4y \ge 7$
$x + 3y \le 4$

10. $y + 1 < 2(x - 3)$
$y - 3 \ge -\frac{1}{2}(x + 2)$

Match each system of linear inequalities with its graph.

11. $y \ge x$
$y \le -x$

12. $y \le x$
$y \ge -x$

13. $y \le x$
$y \le -x$

14. $y \ge x$
$y \ge -x$

A.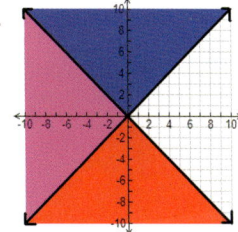
B.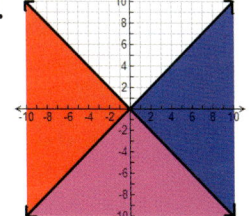
C.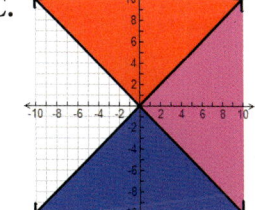
D.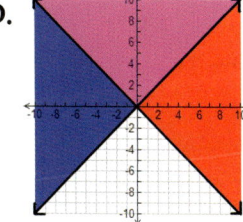

Graph the solution of each system of three linear inequalities.

15. $y < 2x$
$y > 2x - 4$
$x \geq 1$

16. $y \leq 3$
$y \geq -1$
$x < 0$

17. $y > -\frac{1}{2}x + 1$
$y \leq 2x + 1$
$y \geq 1$

18. Is it possible to have a system of two linear inequalities that has no solutions? Support your answer with a graph as an example or counter-example.

19. Sydney wants to write a system of two inequalities that would cause the entire first quadrant to be the solution set. What would her system of inequalities need to be? Use words and/or numbers to show how you determined your answer.

20. Write a system of linear inequalities that is represented in the graph at the right. Use words and/or numbers to show how you determined your answer.

21. Write a system of linear inequalities that contains the points (4, 3), (5, −1), (10, −8) and (0, 0) in the solution set. Explain how you know your answer is correct.

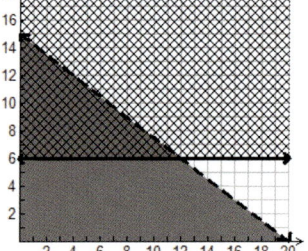

22. Is it possible to have a solution set for a system of linear inequalities that contains points in every quadrant? Explain your reasoning. Use examples or counter-examples, if necessary.

23. What is the area of the region contained by the system of four inequalities below? Show all work necessary to justify your answer.

$y \geq -2$ $\qquad x \geq 0$
$y \leq 5$ $\qquad x \leq 3$

REVIEW

Decide whether the given ordered pair is a solution of the system of equations. Show all work necessary to justify your answer.

24. $y = 3x$
$y = \frac{1}{2}x + 7$
$(-2, -6)$

25. $x + 3y = 4$
$4x - 5y = -1$
$(1, 1)$

26. $y = x - 7$
$x + 2y = 14$
$(7, 0)$

Solve each system of equations by graphing. Show all work necessary to prove that your answer is correct. If there is not exactly one solution, determine if the system of equations has infinitely many solutions or no solution.

27. $y = x - 3$
$y = 2x + 3$

28. $y = \frac{2}{5}x - 2$
$y = -x + 5$

29. $y = \frac{1}{2}x - 1$
$y = 2$

30. $y = -\frac{1}{2}x + 1$
$x + 2y = 2$

31. $y = 1 - \frac{2}{3}x$
$y = \frac{1}{6}x - 4$

32. $-2x + 4y = 8$
$y = \frac{1}{2}x - 3$

CONVERTING REPEATING DECIMALS TO FRACTIONS

LESSON 4.9

 Convert repeating decimals to fractions.

A number that can be expressed as a fraction of two integers is called a **rational number**. Every rational number can be written as a decimal number. The decimal numbers will either terminate (end) or repeat.

Terminating Decimals			Repeating Decimals		
$\frac{3}{4} = 0.75$	$\frac{19}{10} = 1.9$	$\frac{394}{1000} = 0.394$	8.16161616... is $8.\overline{16}$	0.3333... is $0.\overline{3}$	5.0424242... is $5.0\overline{42}$

The bar goes above the 16 because both digits repeat.

The bar goes above the 3 because it is the number that repeats.

The bar goes only above the 42 because they are the only two digits that repeat.

Converting a repeating decimal to a fraction can be done by creating an equation or system of equations and then solving those equations.

CONVERTING A REPEATING DECIMAL TO A FRACTION

1. Let x equal the repeating portion of the decimal number.
2. Multiply both sides of the equation by a power of ten to move the repeating digit(s) to the left side of the decimal point.
3. Subtract x from both sides of the equation.
4. Solve for x.

EXAMPLE 1   Convert $0.\overline{4}$ to a fraction.

SOLUTION

Let x equal the repeating decimal. $x = 0.\overline{4}$ or $x = 0.4444...$

Multiply both sides of the equation by 10. This moves the repeating digit to the left side of the decimal point.
$10x = 10(0.4444...)$
$10x = 4.444...$

Remember that x is equal to 0.444...

Subtract x from both sides of the equation.
$10x = 4.444...$
$-x\ \ -0.444...$
$9x = 4$
$x = \frac{4}{9}$

$0.\overline{4} = \frac{4}{9}$

$\frac{4}{9}$ is the fraction equal to $0.\overline{4}$.

Lesson 4.9 ~ Converting Repeating Decimals to Fractions   **171**

EXAMPLE 2   David had $1.\overline{18}$ ounces of silver. What is this amount as a fraction?

SOLUTION

Let x equal the repeating decimal. $x = 0.1818...$

Ignore the whole number while finding the fraction.

Multiply both sides of the equation by 100. This moves the repeating digit to the left side of the decimal point.

$100x = 100(0.1818...)$
$100x = 18.1818...$

Remember that x is equal to $0.1818...$

Subtract x from both sides of the equation.

$$100x = 18.1818...$$
$$\underline{-x \quad -0.1818...}$$
$$99x = 18$$

Now that the fraction is found, add the whole number back in.

Divide both sides of the equation by 99. $x = \dfrac{18}{99} \rightarrow 1\dfrac{18}{99}$

$1.\overline{18}$ ounces of silver is equivalent to $1\dfrac{18}{99}$ ounces.

When the repeating decimal digits fall after digits that do not repeat, you need to set up a system of equations to find the equivalent fractions.

EXAMPLE 3   What is $0.8\overline{3}$ as a fraction?

SOLUTION

Let x equal the repeating decimal. $x = 0.8\overline{3}$ or $x = 0.8333...$

Multiply both sides of the equation by 100. This moves the repeating digit (3) to the left side of the decimal point.

$100x = 100(0.8333...)$
$100x = 83.333...$

Any other power of 10 will work.

Create another equation by multiplying both sides of the original equation by a different power of 10. This will allow you to subtract the repeating part of the decimal.

$10x = 8.3333...$

Subtract the equations from one another.

$$100x = 83.3333...$$
$$\underline{-10x \quad -8.3333...}$$
$$90x = 75$$

Divide both sides of the equation by 90. $x = \dfrac{75}{90} = \dfrac{5}{6}$

$0.8\overline{3} = \dfrac{5}{6}$

EXERCISES

Label each of the following decimals as a "terminating decimal" or a "repeating decimal".

1. 5.45 **2.** 7.$\overline{6}$ **3.** 2.08$\overline{4}$

4. 6.32 **5.** 0.8$\overline{58}$ **6.** 23.769

Match the following repeating decimals with its equivalent fraction value in the box.

7. 0.$\overline{18}$ **8.** 0.$\overline{6}$

9. 0.1$\overline{6}$ **10.** 0.41$\overline{6}$

11. 0.$\overline{1}$ **12.** 0.6$\overline{3}$

Rational Numbers

$\frac{2}{3}$ $\frac{1}{9}$ $\frac{1}{6}$ $\frac{5}{12}$ $\frac{2}{11}$ $\frac{19}{30}$

Convert each repeating decimal into a fraction.

13. 0.$\overline{2}$ **14.** 0.$\overline{5}$ **15.** 0.$\overline{15}$

16. 0.$\overline{63}$ **17.** 0.1$\overline{2}$ **18.** 0.04$\overline{3}$

19. 0.$\overline{414}$ **20.** 0.36$\overline{3}$ **21.** 0.$\overline{162}$

22. In what situations must you set up multiple equations when converting a repeating decimal to a fraction? Write an example of one such type of decimal number.

23. Why are powers of 10 chosen as the multipliers for converting decimals to fractions?

24. Hank has a snake which weighs 3.$\overline{7}$ ounces. Jill has a lizard which weighs 3$\frac{4}{5}$ ounces. Whose reptile weighs more? Support your solution with calculations.

25. Megan ran a mile in 9.$\overline{45}$ minutes. Janeen ran a mile in 9$\frac{4}{9}$ minutes. Anna ran a mile in 9$\frac{13}{30}$ minutes.
  a. Put the runners' times in order from fastest to slowest.
  b. Who was the fastest?

26. When a single digit repeats after the decimal point (i.e., 0.$\overline{1}$, 0.$\overline{2}$, 0.$\overline{3}$, 0.$\overline{4}$, etc), what do you notice about the denominators of the equivalent fractions? Use words and/or numbers to show how you determined your answer.

27. Find a repeating decimal that has a value between $\frac{3}{4}$ and $\frac{7}{8}$. Write the value as a fraction and a decimal. Use words and/or numbers to show how you determined your answer.

28. Find a repeating decimal that has a value between 1$\frac{1}{4}$ and 1$\frac{1}{3}$. Write the value as a fraction and a decimal. Use words and/or numbers to show how you determined your answer.

Lesson 4.9 ~ Converting Repeating Decimals to Fractions 173

REVIEW

29. Two beach house rentals have different pricing systems. House A charges a flat fee for cleaning of $90 plus $120 per night. House B does not have a cleaning fee but charges $142.50 per night.
  a. How many nights would you need to stay so that both houses charged the same amount?
  b. How much would it cost to rent one of the houses after this many nights?

30. Julie has a total of 23 granola bars and candy bars. The granola bars have 90 calories each. The candy bars each have 240 calories. Together her treats have a total of 2,970 calories. How many of each type of treat does she have? Show all work necessary to justify your answer.

31. Jacob and Jordan each decided to buy movie tickets and popcorn with their monthly allowance. Jacob bought 5 movie tickets and 3 popcorns for $63. Jordan bought 9 movie tickets and 2 popcorns for $93.
  a. What is the cost for a single movie ticket?
  b. What is the cost for a single popcorn?
  c. Kylie was at the same theater and bought 4 movie tickets and 3 popcorns for her family. How much did she spend? Show all work necessary to justify your answer.

32. Solve the system of equations using two different methods.
$$2x + 3y = 2$$
$$x - 3y = 10$$

Tic-Tac-Toe ~ "How To" Guide

Design a brochure that explains in detail how to solve a system of equations using at least two different methods. Include multiple examples and step-by-step instructions. Include diagrams and helpful hints, when necessary.

Tic-Tac-Toe ~ Pros and Cons

Four methods for solving systems of linear equations have been introduced in this block. The methods include graphing, input-output tables, elimination and substitution. Create a visual display showing the positive and negative aspects of each method. Describe how a system might be set up in a way that would make one method easier than another. Include an example for each method.

REVIEW BLOCK 4

Vocabulary

elimination method
parallel
rational numbers

solution to a system of linear equations
substitution method
system of linear equations

Algebraically determine if two lines are parallel, intersecting or the same line.
Determine the solution to a system of equations by graphing.
Determine the solution to a system of equations using tables.
Determine the solution to a system of equations using the substitution method.
Determine the solution to a system of equations using the elimination method.
Choose the best method for solving a given system of equations.
Set up and solve systems of equations from word problems.
Solve a system of linear inequalities by graphing.
Convert repeating decimals to fractions.

Lesson 4.1 ~ Parallel, Intersecting or The Same Line

Determine if each graph shows a system of linear equations that is intersecting, parallel or the same line. State how many solutions there are for each system.

1. 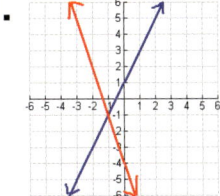 **2.** 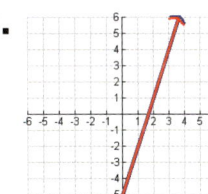 **3.**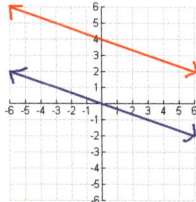

Algebraically determine if the two lines in each system of equations are intersecting, parallel or the same line. State how many solutions there will be for each system. Use words and/or numbers to show how you determined your answer.

4. $y = -\frac{1}{2}x - 4$
$y = \frac{1}{2}x + 4$

5. $3x + 4y = 20$
$y = -\frac{3}{4}x - 2$

6. $-6x + 3y = 12$
$y = 2(x + 1) + 2$

Lesson 4.2 ~ Solving Systems by Graphing

Decide whether the given ordered pair is a solution of the system of equations. Show all work necessary to justify your answer.

7. $y = 2x - 1$
$y = \frac{1}{3}x + 4$
$(3, 5)$

8. $x - 3y = 4$
$4x + y = -5$
$(-1, 1)$

9. $y = x - 2$
$4x + y = -2$
$(0, -2)$

Solve each system of equations by graphing. Show all work necessary to prove that your answer is correct.

10. $y = \frac{1}{2}x - 3$
$y = x - 5$

11. $y = -2x - 4$
$y = \frac{1}{2}x + 1$

12. $y = -\frac{1}{3}x$
$y = 1$

13. Sandra and Terry each walk to the same school from different neighborhoods. They do not cross paths until they reach the school building. Sandra follows the path represented by the equation $y = -2x + 10$ and Terry follows the path represented by the equation $y = -\frac{1}{6}x - 1$. What are the coordinates of the school building?

Lesson 4.3 ~ Solving Systems Using Tables

Solve each system of equations using the given input-output tables. Show all work necessary to prove that your answer is correct.

14.

$y = 5x - 4$

x	y
0	
1	
2	
3	

$y = -2x + 3$

x	y
0	
1	
2	
3	

15.

$y = -5 + x$

x	y
-5	
-4	
-3	
-2	

$y = 3x + 3$

x	y
-5	
-4	
-3	
-2	

Solve each system of equations using input-output tables. Show all work necessary to prove that your answer is correct.

16. $y = 3x - 4$
$y = x + 2$

17. $y = -4x + 1$
$y = 2x + 13$

18. $y = x$
$y = 3x - 8$

19. Evan had $400 in a savings account at the beginning of the year. At the end of each month, he took $20 out of the account. Lisa put $100 in her savings account at the beginning of the year. At the end of each month, she put $40 in her account. Let x represent the number of months which have passed and y represent the amount in each savings account.
 a. Write an equation to represent the amount in Evan's savings account.
 b. Write an equation to represent the amount in Lisa's savings account.
 c. Copy and complete the input-output tables through 6 months.

Evan's Savings Account Balance

Months, x	Total Savings, y
0	
1	
2	

Lisa's Savings Account Balance

Months, x	Total Savings, y
0	
1	
2	

 d. When will Evan and Lisa have the same amount in their savings accounts? How much will they each have at this time?

Lesson 4.4 ~ Solving Systems by Substitution

Solve the system of equations using the substitution method. Show all work necessary to prove that your answer is correct.

20. $x = y + 1$
$3x + 2y = 18$

21. $-3x + y = 9$
$y = 2x + 6$

22. $x + y = 3$
$x + 2y = 1$

23. Both of Monique's neighbors owned cows. Mr. James owned five less than three times the number of cows owned by Mr. Peters. The total number of cows owned by both neighbors was 79. Let x represent the number of cows Mr. James owns and y represent the number of cows Mr. Peters owns.
  a. Explain why the equations $x = 3y - 5$ and $x + y = 79$ represent this situation.
  b. Solve the system of equations using the substitution method. How many cows did each neighbor own?

Lesson 4.5 ~ Solving Systems Using Elimination

Show all work necessary to transform each system of linear equations into a system that is ready for columns to be added together to eliminate a variable. Rewrite the system if it is already set up for elimination.

24. $-x + y = 12$
$x + y = 6$

25. $3x - y = 3$
$6x + 2y = 9$

26. $4x + 2y = 9$
$-2x + 3y = 1$

Solve each system of equations using the elimination method. Show all work necessary to prove that your answer is correct.

27. $x + 3y = 17$
$2x - 3y = -20$

28. $5x + 4y = 22$
$2x + 4y = 16$

29. $2x + y = 7$
$4x - 3y = -6$

30. Patrick bought one baseball cap and one t-shirt for $36. Sammy bought two baseball caps identical to Patrick's caps along with three of the same t-shirts. Sammy spent a total of $94.
  a. Explain why the equations $x + y = 36$ and $2x + 3y = 94$ represent this situation.
  b. What does x represent based on the equations in **part a**? What does y represent?
  c. Solve the system of equations using the elimination method. What are the individual costs for a t-shirt and a baseball cap?

Lesson 4.6 ~ Choosing the Best Method

State the best method to solve each system of linear equations and then solve.

31. $2x + y = 6$
$x - y = 6$

32. $x = y - 1$
$3x + y = 13$

33. $y = \frac{1}{3}x - 5$
$y = -\frac{4}{3}x$

34. $3x + 4y = 7$
$x - 8y = 0$

35. $y = \frac{1}{2}x + 3$
$-3x + 4y = 18$

36. $y = 2x - 5$
$4x + y = -11$

Lesson 4.7 ~ Applications of Systems of Equations

Solve each problem using a system of equations. Define the variables and state the solution. Explain how you know your answer is correct.

37. Two jet ski rental companies have different costs. Company A charges a flat fee of $8 plus $2.50 per hour. Company B charges a flat fee of $14 plus $1.00 per hour. At what point in time are both rentals the same amount? How much are the rentals for that amount of time?

38. The Mendenhall Theater sells two types of tickets: youth and adult. The theater holds a total of 450 people. One night, the theater sold all their tickets for a total of $2,706. Youth tickets cost $4.60 and adult tickets cost $7.00. How many tickets of each type did the theater sell that night?

39. Jamal and Emily each started a savings account in January. Jamal started with $46 in his account and added $24 each month. Emily opened her account with $319. Each month she withdrew $15. After how many months will they have the exact same amount in their accounts? How much will be in their accounts at that time?

40. Two girls sold lemonade together. The entire lemonade sale brought in $28. One girl made $4 more than twice the amount the second girl made. How much did each girl make at the lemonade sale? Explain how you know your answer is correct.

Lesson 4.8 ~ Systems of Linear Inequalities

Graph the solution of each system of linear inequalities.

41. $y > -3x + 4$
$y < \frac{1}{3}x - 6$

42. $y \leq x + 1$
$y > 5 - 2x$

43. $y \leq \frac{2}{3}x$
$y \geq \frac{2}{3}x - 4$

44. $y > 2$
$x > -2$

45. $y \geq \frac{1}{4}x - 5$
$y \geq -\frac{5}{4}x + 1$

46. $y < 3 - 4x$
$y \leq 2x - 4$

47. Yolanda graphed the system of inequalities where $y \geq -1$ and $y \leq 1$. She said the answers to the system of inequalities was in the middle purple region of the graph below. Do you agree or disagree? Explain your reasoning.

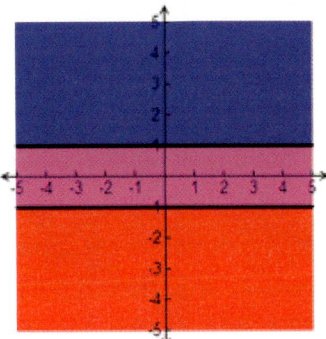

Lesson 4.9 ~ Converting Repeating Decimals to Fractions

Label each of the following decimals as a "terminating decimal" or a "repeating decimal".

48. 0.04

49. $1.\overline{3}$

50. 0.5656

Convert each repeating decimal into a fraction.

51. $0.\overline{5}$

52. $2.\overline{6}$

53. $0.\overline{23}$

54. $0.\overline{81}$

55. $1.5\overline{3}$

56. $7.0\overline{7}$

57. Find a repeating decimal that has a value between $\frac{1}{2}$ and $\frac{3}{5}$. Write the value as a fraction and a decimal. Use words and/or numbers to show how you determined your answer.

CAREER FOCUS

RYAN
ACCOUNTANT

My name is Ryan and I am an accountant. Accountants track financial information for businesses and other organizations. This can mean keeping track of data, preparing reports or using numbers to make predictions. Managers and leaders of businesses depend on accountants to give them accurate information so that they can make decisions that best benefit their organization. Accountants also help organizations with figuring out how much money they will need to pay in taxes.

I use math in many ways. I constantly use basic operations to transform raw data into usable information. I have to problem solve in different situations to analyze data for correctness. I also use ratios and fractions to compare how one business is doing as opposed to another. There are a number of types of statistics I use to prepare forecasts and valuations.

A Certified Public Accountant must have a 4-5 year degree in accounting. They must also pass a national exam and have a certain number of hours of on-the-job experience in order to get a license. The on-the-job experience can be done while working under the supervision of a CPA who already has a license.

The starting salary for an accountant is around $36,000 per year. Average salaries range between $43,000 and $67,000 per year depending on the industry and experience level of the accountant. Salaries for accountants can sometimes reach $150,000 per year or more.

I like the many kinds of organizations and industries an accountant can work in. An accountant can work in government, charity or business. They can be entrepreneurs and open their own public accounting firm, or they can work for a Fortune 500 company. They can work in the entertainment industry or in agriculture. Another benefit to working in accounting is always being a key player in the important decisions that shape and direct the organization you work for.

CORE FOCUS ON LINEAR EQUATIONS
BLOCK 5 ~ TWO-VARIABLE DATA

Lesson 5.1	Scatter Plots and Correlation	183
Lesson 5.2	Predicting with Lines of Best Fit	187
	Explore! Finding a Good Fit	
Lesson 5.3	Five-Number Summaries of Data	192
Lesson 5.4	Q-Points and Lines of Best Fit	197
	Explore! The Wave	
Lesson 5.5	Predicting with Best Fit Equations	203
Lesson 5.6	Using Data and Graphs to Persuade	207
	Explore! Eliminating Bias	
Lesson 5.7	Bivariate Data and Frequency Tables	212
Review	Block 5 ~ Two-Variable Data	219

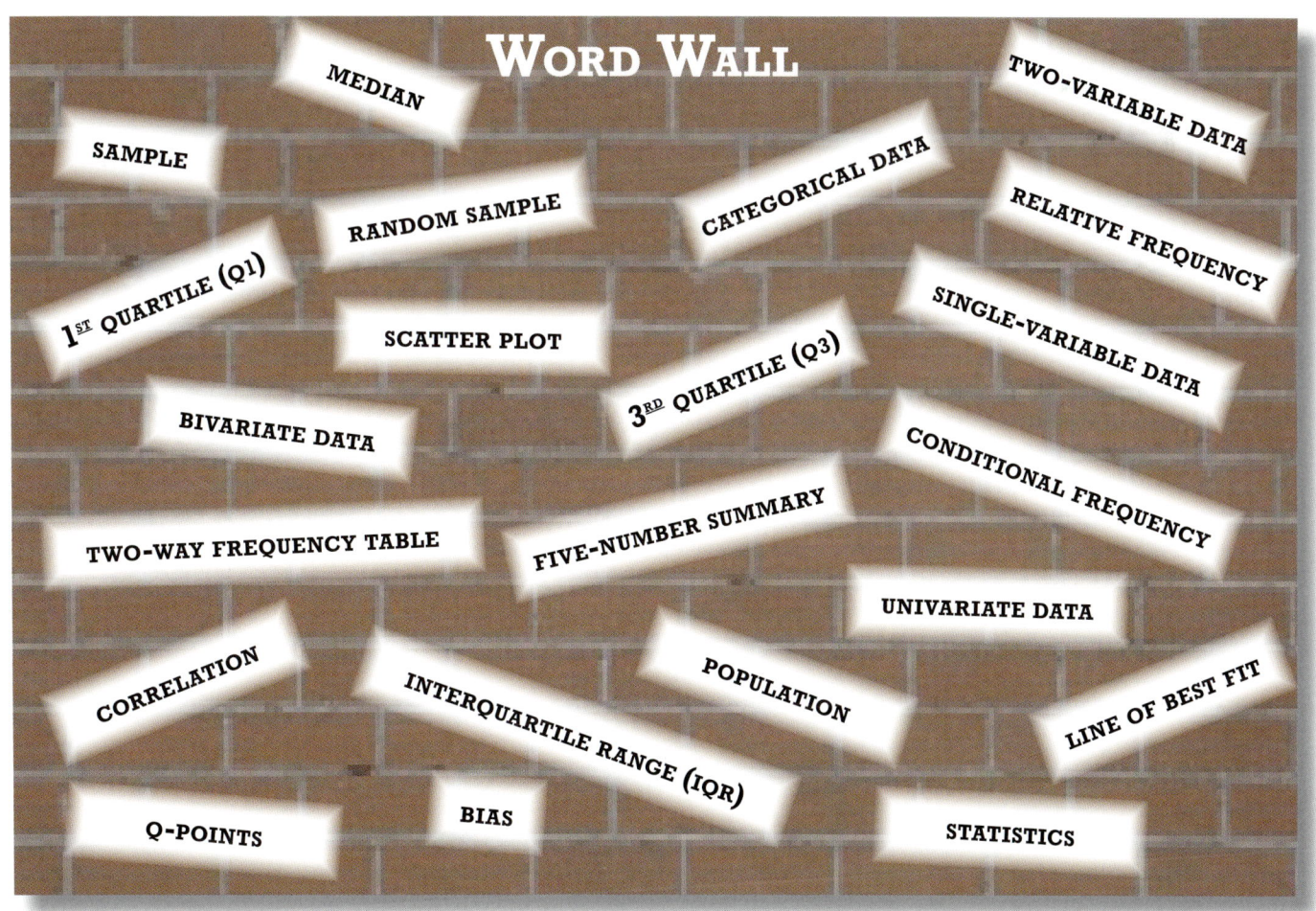

BLOCK 5 ~ TWO-VARIABLE DATA
TIC-TAC-TOE

PREDICTION TUTORIAL

Make a brochure to teach a new student how to use scatter plots or lines of best fit to make predictions.

See page 191 for details.

TEST A THEORY

Make conjectures about the relationship between two variables. Collect data to see if the theories are correct.

See page 196 for details.

CORRELATION COEFFICIENTS

Research correlation coefficients and create a booklet describing your findings.

See page 196 for details.

CONDUCT A BIVARIATE SURVEY

Create a short survey to test whether there is a relationship between two sets of data.

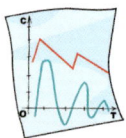

See page 218 for details.

GLOSSARY FLASHCARDS

Create flashcards of all vocabulary terms in this block.

See page 223 for details.

PATTERNS IN SPORTS

Explore the correlations and variables in a sport. Describe how the results could help a coach make team decisions.

See page 191 for details.

CORRELATIONS IN THE REAL WORLD

Find examples of positive and negative correlations in newspapers or magazines.

See page 218 for details.

Q-POINTS POSTER

Create a poster that summarizes the process of finding a line of best fit by Q-points.

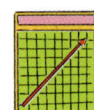

See page 211 for details.

GOOD SURVEY/BAD SURVEY

Create a survey with good questions. Change the questions to make the same survey a bad survey.

See page 211 for details.

SCATTER PLOTS AND CORRELATION

LESSON 5.1

 Read, create and describe the correlations in scatter plots.

Is there a relationship between people's education level and their salaries? Is there a relationship between the number of missing assignments that students have and their grades?

Statistics is the process of collecting, displaying and analyzing a set of data. Consumers may use data to decide which products to buy. Businesses use data to make important financial decisions. In the past you have focused on **single-variable data**. These data sets only showed one type of data, perhaps the weights of frogs or the heights of buildings. In this block, the focus will shift to **two-variable data**, where two types of data will be considered at the same time.

It is often helpful to plot two-variable data sets as ordered pairs. Plotting the points from a two-variable data set can help you see if there is a relationship between the two variables. This set of points is called a **scatter plot**.

Year	Time (seconds)
1956	11.5
1972	11.07
2004	10.93
2008	10.78
1996	10.94
1988	10.54
1936	11.5
1980	11.06
1948	11.9
2000	10.75
1964	11.4
1984	10.97
1992	10.82
2012	10.75

Source: Time Almanac

The data at right shows the times from the Women's 100-Meter Dash from the last several Summer Olympic Games. The *x*-values vary from 1936 to 2012. The *y*-values vary from 10.54 to 11.9 seconds. The axes have been scaled to include these values. Each point has been plotted where (*x*, *y*) represents (Year, Time in seconds).

The relationship between the variables can be shown by plotting the points. The graph shows a downward pattern. The pattern shows that, in general, as the years increase, the women's times decrease. Women's Olympic runners are getting faster over time.

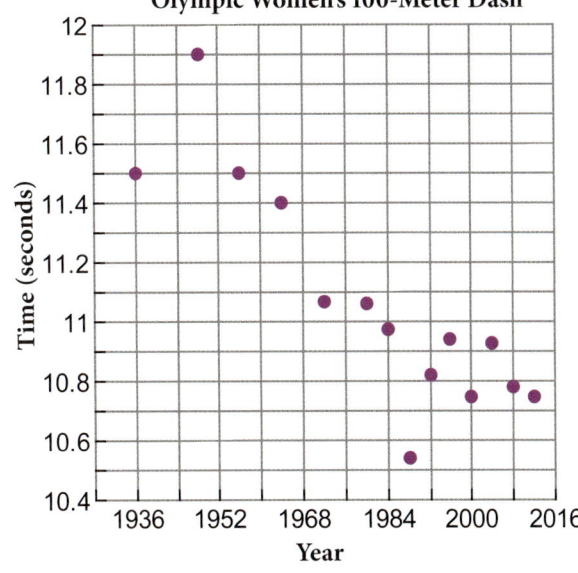

Lesson 5.1 ~ Scatter Plots and Correlation   183

EXAMPLE 1

Make a scatter plot of this data set. Describe the pattern of the data.

x	1	3	5	2	8	4	5	4	2	6	7
y	3	3	7	2	8	5	5	6	4	6	8

SOLUTION

The *x*-values vary from 1 to 8. The *y*-values vary from 2 to 8. An appropriate scale that includes all ordered pairs is 0-8 on the *x*- and *y*-axes.

Plot the points. In general, as the *x*-values increase, the *y*-values also increase.

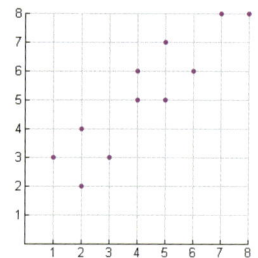

A scatter plot can be very helpful in determining if there is a relationship between a pair of variables in a data set. A **correlation** describes the relationship between the two variables in a scatter plot. There are three types of correlations.

Positive Correlation
Studying and Test Scores

Negative Correlation
Car's Value Over Time

No Correlation
Weekly Grocery Expenses by Height

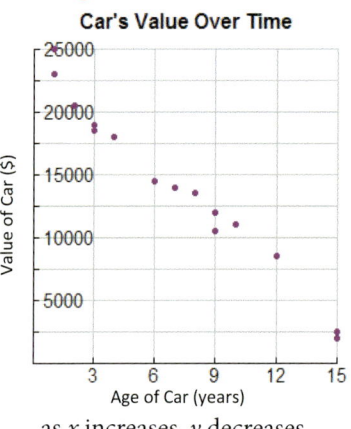
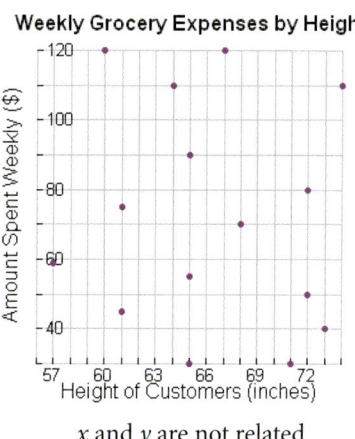

as *x* increases, *y* increases   as *x* increases, *y* decreases   *x* and *y* are not related

Notice that correlations are not always perfect patterns. A correlation shows the general pattern or trend of the data. When there is no correlation, the ordered pairs in the graph do not show any recognizable pattern or trend.

EXAMPLE 2

Predict whether the following sets of data would show positive, negative or no correlation. Explain.
a. a student's grade level and the amount of recess time
b. a student's height and their score on the last science test
c. a student's height and their shoe size

SOLUTIONS

a. As students get older, the amount of time they get for recess typically decreases. This would likely show a negative correlation.

b. The height of a student should have no effect on how they performed on the last science test. There is likely no correlation.

c. Shorter students are more likely to have smaller feet and taller students are more likely to have larger feet. This is a positive correlation.

EXERCISES

1. How are scatter plots related to ordered pairs?

2. How is a scatter plot used to determine if two-variable data shows no correlation?

3. Draw a coordinate plane and label the *x*- and *y*-axes. Plot and label the following ordered pairs.

A (6, 1)   **B** (0, −5)   **C** (−3, 7)   **D** (2, 8)
E (3, 0)   **F** (1, −8)   **G** (−4, −1)   **H** (7, −2)

4. Write the ordered pair for each point in the scatter plot at right.

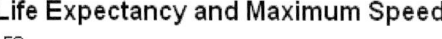

5. The scatter plot at right shows the life expectancy and maximum speed of several animals. Use the plot to answer the following questions.
   a. Which animal has the fastest maximum speed?
   b. Which animal has the greatest life expectancy?
   c. Which animal has the lowest life expectancy?
   d. How do an elephant and a grizzly bear compare in terms of their life expectancy and maximum speed?
   e. Does there appear to be a correlation between life expectancy and maximum speed of animals? Explain your reasoning.

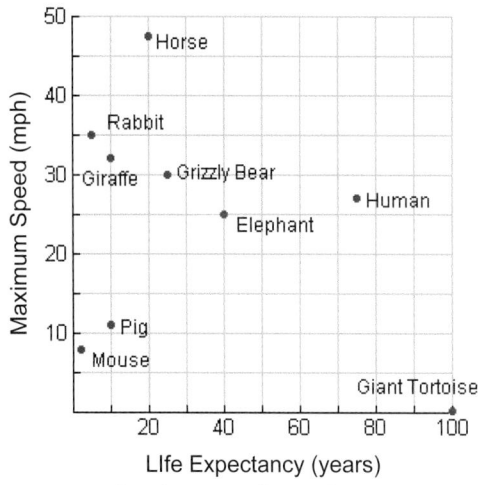

Predict whether each set of data would show positive, negative or no correlation. Explain each answer.

6. a toad's weight and its jumping distance

7. price a gas station charges per gallon and the number of customers they have per day

8. time spent studying and the score received on a test

9. time spent brushing teeth and the number of cavities

10. Hayden wondered if there was a relationship between the minutes students spent exercising each day and the minutes they spent doing homework. He asked several of his classmates to keep track of each for one day. He recorded the data in the table at right.

    a. Make a scatter plot of the data. Let x represent the minutes spent exercising. Let y represent the number of minutes spent on homework.
    b. Which two points seem out of place on the scatter plot? Explain your reasoning.
    c. Aside from these two points, describe the correlation.
    d. Hayden was surprised by the correlation he found. Why might the time spent exercising actually lead to more time spent on homework?

Minutes Spent Exercising	Minutes Spent on Homework
25	30
45	70
40	50
5	80
15	40
30	50
15	15
60	20
35	40
40	60
50	60
20	40

Movie	Year Released	Revenue (in millions)
Avatar	2009	2,782
Titanic	1997	2,185
The Avengers	2012	1,482
Harry Potter and the Deathly Hallows, Part II	2011	1,328
Transformers: Dark of the Moon	2011	1,124
The Lord of the Rings: The Return of the King	2003	1,120
Pirates of the Caribbean: Dead Man's Chest	2006	1,066
Toy Story 3	2010	1,063
Pirates of the Caribbean: On Stranger Tides	2011	1,044
Star Wars: Episode 1 - The Phantom Menace	1999	1,027
Alice in Wonderland	2010	1,024

Source: Time Almanac

11. As of 2012, "Avatar" was the highest-grossing film ever. The table at left shows the revenue of the all-time top movies as well as the year they were released.

    a. Make a scatter plot of the data in the table to show the year released (x) and the revenue in millions (y).
    b. Is there a correlation between the year the movie was released and its revenue? Explain.

REVIEW

Solve each system of equations. Show all work necessary to justify each answer.

12. $y = 5x + 8$
    $2x - 3y = 15$

13. $-4x + 12 = 2y$
    $3x + 5y = 23$

14. $2x - y = 3$
    $3x + 2y = 36$

Solve each problem using a system of equations. Define the variables and state the solution.

15. Geno is building a rectangular pen for his chickens. The design of the pen calls for the length to be 4 more than twice the width of the pen. He has 104 feet of fencing. What is the maximum width and length possible for the chicken pen?

16. Claudia went to the concession stand at the baseball game and bought 3 hot dogs and 4 drinks. Her total came to $18.50. Marco spent $18 for 4 hot dogs and 2 drinks. How much does a hot dog cost at this concession stand? How much does a drink cost?

PREDICTING WITH LINES OF BEST FIT

LESSON 5.2

 Draw a line of best fit for a set of data.
Use a line of best fit to make predictions.

In **Lesson 5.1**, you learned how to create and interpret scatter plots. A scatter plot often reveals a recognizable pattern in a data set. For example, sometimes positively or negatively correlated scatter plots form a linear pattern. When this occurs, it can be helpful to determine a **line of best fit** for the data. A line of best fit approximates the pattern of the data. It is useful for making predictions about the data set.

There are two criteria for a good line of best fit. First, the line must follow the direction of the linear data well. Also, the line should have about half the points above the line and half the points below the line.

EXPLORE!   FINDING A GOOD FIT

Refer to the scatter plots below:

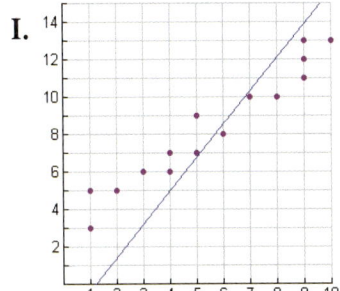
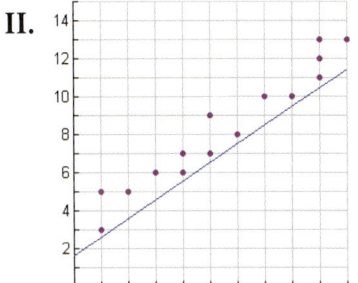
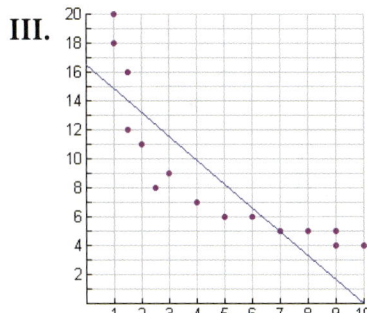
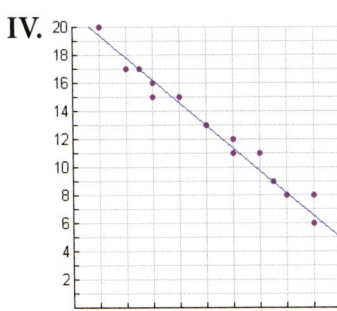
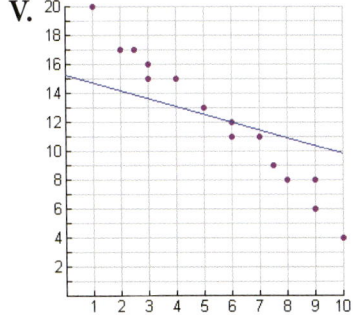
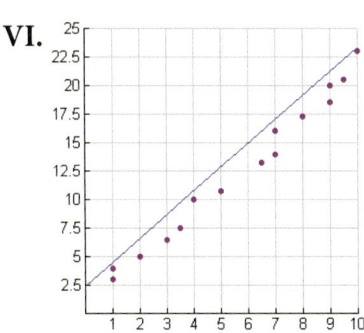

Step 1: Which scatter plots show a positive correlation? Which show a negative correlation?

Step 2: Which scatter plots follow a linear pattern?

Step 3: Which scatter plot has a good line of best fit drawn?

Step 4: Which scatter plots do not have a good line of best fit drawn? Explain what is wrong with each.

Lesson 5.2 ~ Predicting with Lines of Best Fit   **187**

LINE OF BEST FIT

A good line of best fit:
- follows the direction of a linear pattern, and
- has approximately half the points above and half the points below the line.

EXAMPLE 1

Use the scatter plot of Federal Minimum Wage by Year to complete the following.

a. Draw a line of best fit for the scatter plot. Justify why the line of fit is good.
b. Use the line of best fit to predict what the federal minimum wage will be in the year 2020.

SOLUTIONS

a. A good line of best fit will follow the pattern of the data and will have about half the points above and below the line. This line is a good approximation.

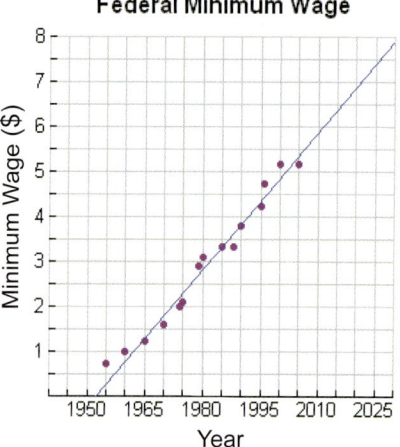

b. Extend the line of best fit as well as the x- and y-axes. Find the year 2020 on the x-axis (Years) and follow it up to where it meets the line of best fit. Now follow that point to the y-axis (Minimum Wage in Dollars).

Locate the y-value that matches with the year 2020 on the line. This represents the approximate minimum wage. It appears that the minimum wage will be about $6.75 in 2020.

EXERCISES

1. What are the characteristics of a good line of best fit?

Identify whether each graph has a good line of best fit drawn. If the line is not a good fit, explain why.

2.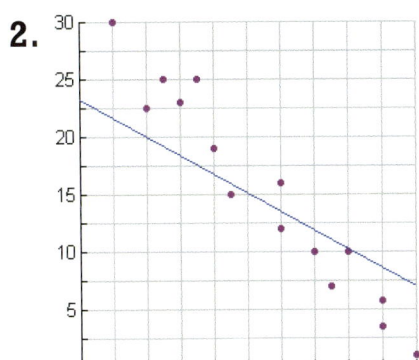

3.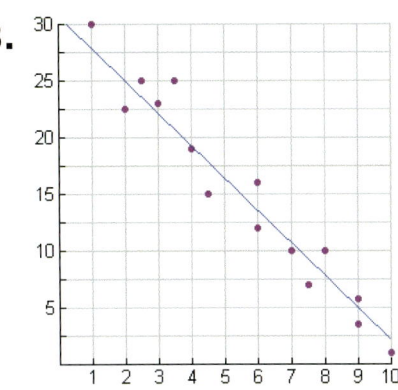

4.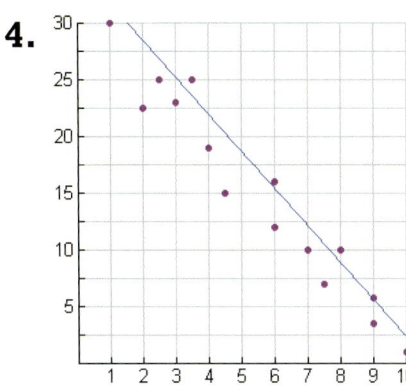

5. Dave drew a line of best fit for the scatter plot at right.
  a. How many points are above Dave's line of best fit?
  b. How many points are below Dave's line of best fit?
  c. Scott said Dave's line does not fit the pattern of the data very well. Do you agree or disagree? Explain your reasoning.
  d. Should the data in the scatter plot be represented by a line of best fit? Why or why not?

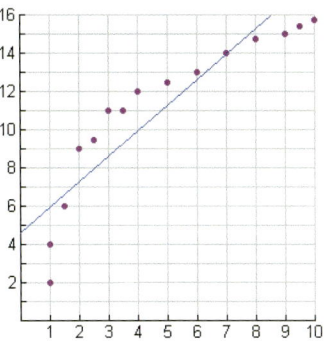

6. Inessa plotted her age and height from birth to when she turned 13 on a scatter plot. She noticed that as she got older, she also got taller. She wants to draw a line of best fit through the data to predict how tall she will be when she is 30. Explain why her line of best fit will not work in this situation.

Use the scatter plot and line of best fit to approximate each missing value.

7. When $x = 14$, $y = $ _____

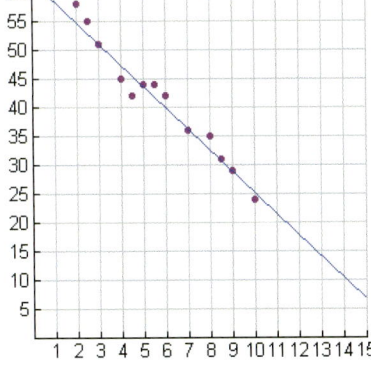

8. When $y = 38$, $x = $ _____

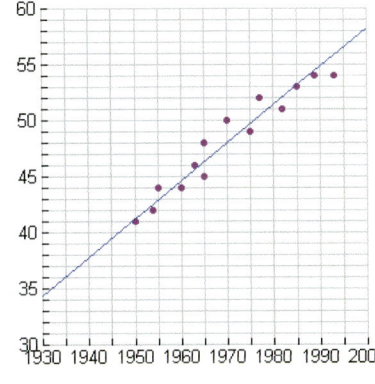

9. When $x = 26$, $y = $ _____

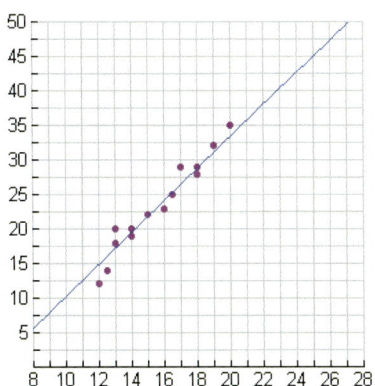

Lesson 5.2 ~ Predicting with Lines of Best Fit   **189**

10. Morgan wanted to see if there was a correlation between a person's age and the number of hours they watch television each week. He surveyed several people and then put the information into the following table.
    a. Create a scatter plot of the data. Scale the *x*-axis from 0 to 60 years old and the *y*-axis from 10 to 45 hours.
    b. Describe the correlation of the data. Does the data appear to be linear?
    c. Draw a line of best fit for the data. Extend your line all the way through the graph. Based on Morgan's data, how many hours of television would you expect a 50 year old person to watch?
    d. If a person watched 40 hours of television each week, approximately how old would you expect them to be?

Person's Age (x)	Hours Watching TV per Week (y)
13	24
5	23
35	32
25	31
7	18
2	18
14	22
19	24
40	32
26	29
9	20
16	24
22	27
38	34
30	30

11. All the classrooms in a school are to be painted. Several students chose to volunteer for the painting project. The graph below shows the number of students who painted in each room and the area (in square feet) each student painted.
    a. Describe the correlation between the number of students painting in each room and the area each one painted.
    b. Which line of best fit (a or b) on the graph represents the correlation best?
    c. Use the line of best fit you chose in **part b**. How many square feet would you expect each student to paint if they were in a room of 4 students?
    d. If each student painted 1,400 square feet, about how many students were in the room? Use words and/or numbers to show how you determined your answer.

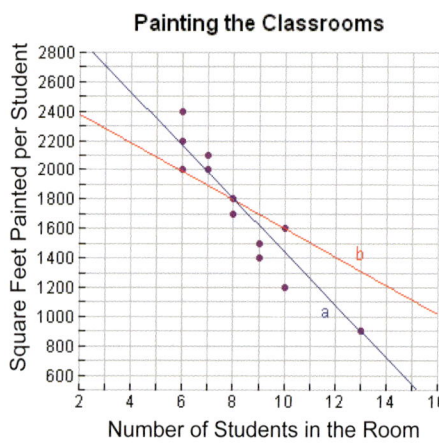

12. Does every scatter plot have a line of best fit? Explain your reasoning and provide an example that supports your answer.

13. Describe a set of data that you could collect from your peers at school that might have a positive correlation. Would a line of best fit for your data be helpful in making predictions? Explain your answer.

REVIEW

14. Sketch an example of a scatter plot with a negative correlation.

Predict whether the following sets of data would show a positive, negative or no correlation.

15. the size of a person's car and the speed they drive

16. the number of traffic tickets a person has received and the cost of their car insurance

17. the number of cars on the highway and the average speed of drivers

18. Write the ordered pair for each point in the scatter plot at the right.

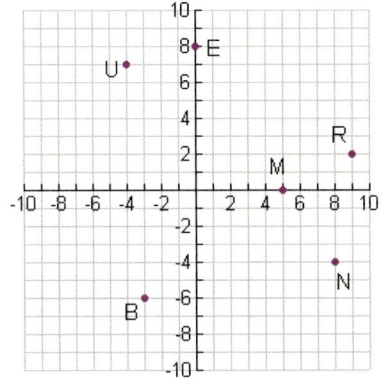

Tic-Tac-Toe ~ Prediction Tutorial

A new student has just joined class. Create a brochure to teach the student how to make predictions using a scatter plot or an equation for a line of best fit. The brochure should explain both processes and give two examples for each. Include some practice problems and their solutions for the new student to try.

Tic-Tac-Toe ~ Patterns in Sports

Are there correlations in sports? Choose one sport and collect data on the relationships among four pairs of variables. For example, for football you could find the relationship between the number of rushing yards per game and the number of wins for various teams as one pair of variables.

Each data set should consist of at least 12 ordered pairs. Create a scatter plot for each of the four pairs of variables. Describe the correlations in each scatter plot. Were the relationships surprising? How could the manager or coach of the team use the scatter plots to help the team? Print or cut out the data used to create the graphs for a visual presentation of the information.

FIVE-NUMBER SUMMARIES OF DATA

LESSON 5.3

 Find the five-number summary of data sets.
Find the interquartile range (IQR) of data sets.

The five-number summary is a helpful statistic that describes the spread of a data set. You will use the **five-number summary** in the next two lessons to help you analyze data.

FIVE-NUMBER SUMMARY

Minimum ~ 1st Quartile (Q1) ~ Median ~ 3rd Quartile (Q3) ~ Maximum

---

Kevin kept track of the total cell phone minutes he used for seven months. Find the five-number summary of the minutes he used by following the steps below.

Kevin: 300, 284, 280, 310, 300, 270, 295

1. Put the numbers in order and find the **median** (the middle number in an ordered data set).	270, 280, 284, 295, 300, 300, 310 Median
2. Find the median of the lower half of the data (this is called the **1st Quartile**).	270, 280, 284, 295, 300, 300, 310 Q1 Median
3. Find the median of the upper half of the data (this is called the **3rd Quartile**).	270, 280, 284, 295, 300, 300, 310 Q1 Median Q3
4. Identify the minimum and maximum values.	270, 280, 284, 295, 300, 300, 310 Min Q1 Median Q3 Max

Do not include the median in the upper or lower half.

The five-number summary of Kevin's cell phone usage data is: 270 ~ 280 ~ 295 ~ 300 ~ 310.

The word "quartile" refers to how the data is separated into quarters. In Kevin's data, 25% of the time he used between 270 and 280 minutes, 25% of the time he used between 280 and 295 minutes, and so on.

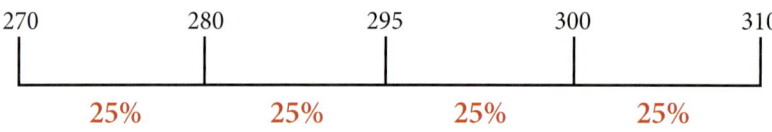

EXAMPLE 1 Find the five-number summary of the data set.
24, 29, 30, 35, 39, 43, 45, 48, 48, 50

SOLUTION

Find the median of the data set.

24, 29, 30, 35, 39, | 43, 45, 48, 48, 50
                41
              median

Find the 1st quartile.
If there are two numbers in the middle, include one in each half of the data.

24, 29, **30**, 35, 39, | 43, 45, 48, 48, 50
       Q1   41

Find the 3rd quartile.

24, 29, **30**, 35, 39, | 43, 45, **48**, 48, 50
                        41   Q3

Find the minimum and maximum.

24, 29, **30**, 35, 39, | 43, 45, **48**, 48, **50**
min   41   max

The five-number summary is 24 ~ 30 ~ 41 ~ 48 ~ 50.

EXAMPLE 2 Find the five-number summary of the data set.
9, 10, 7, 19, 17, 8, 20, 12, 23

SOLUTION

Put the numbers in order.

7, 8, 9, 10, 12, 17, 19, 20, 23

Find the median of the data set.

7, 8, 9, 10, **12**, 17, 19, 20, 23
            median

Find the 1st quartile.
When there is an odd number of values in the set, do not include the median in either half.

7, 8, | 9, 10, **12**, 17, 19, 20, 23
     8.5
      Q1

Find the 3rd quartile.

7, 8, | 9, 10, **12**, 17, 19, | 20, 23
     8.5   19.5
                               Q3

Find the minimum and maximum.

7, 8, | 9, 10, **12**, 17, 19, | 20, **23**
min   8.5   19.5   max

The five-number summary is 7 ~ 8.5 ~ 12 ~ 19.5 ~ 23.

Another statistic which gives a measure of spread is the **interquartile range** or **IQR**. The IQR is the range of the middle half of the data.

INTERQUARTILE RANGE

The interquartile range (IQR) is the difference between the third quartile and the first quartile in a set of data.
IQR = Q3 − Q1

Lesson 5.3 ~ Five-Number Summaries of Data

EXAMPLE 3   The following data lists the average points per game of players on the Miami Heat for the 2011-2012 season. (*Source: espn.com*)

> 3.0, 3.0, 3.4, 3.6, 3.6, 4.8, 6.0, 6.1, 6.8, 9.8, 18.0, 22.1, 27.1

a. Find the five-number summary for the data set.
b. Find the range and interquartile range of the average points per game by the players.

SOLUTIONS

a. The middle number of the data set (the median) is 6.0. The medians of the lower and upper halves, Q1 and Q3, are 3.5 and 13.9, respectively. The minimum is 3.0 and the maximum is 27.1.

$$\underline{3.0}, 3.0, 3.4, | 3.6, 3.6, 4.8, \underline{6.0}, 6.1, 6.8, 9.8, | 18.0, 22.1, \underline{27.1}$$

| min | Q1 (3.5) | Median | Q3 (13.9) | max |

The five-number summary for the data is 3.0 ~ 3.5 ~ 6.0 ~ 13.9 ~ 27.1

b. Find the range of the data.   Range = Maximum − Minimum = 27.1 − 3.0 = 24.1

Find the interquartile range.   IQR = Q3 − Q1 = 13.9 − 3.5 = 10.4

EXERCISES

1. Misty found the five-number summary for the following set of data.
Number of students in each Algebra class:
19, 23, 23, | 25, 26, 26, 27, 27, | 28, 28, 30
19 ~ 24 ~ 26 ~ 27.5 ~ 30

a. Antwan tells Misty her answer is only partly correct. Which part(s) did she have right?
b. Which parts did she get wrong? Fix her errors and give the correct five-number summary.
c. What is the range of the class sizes? What is the interquartile range (IQR)?
d. About ___% of Algebra classes have 23 or fewer students.
e. One-half of the Algebra classes have more than ___ students.

Find the five-number summary for each data set.

2. 7, 14, 15, 19, 20, 25, 27

3. 40, 43, 45, 47, 54, 60

4. 12, 15, 12, 20, 16, 12, 8, 19, 18, 5

5. 74, 76, 79, 68, 52, 59, 76, 60, 78

6. 20, 22, 22, 23, 27, 32, 35, 40, 42, 45, 48, 50

7. −5, 1, 0, 5, 0, 10, −6, −5, 9, 7, −9

Given the five-number summaries, find the interquartile range (IQR) for each data set.

8. 21 ~ 23 ~ 24 ~ 29 ~ 35

9. 16 ~ 28 ~ 34 ~ 37 ~ 40

10. 67 ~ 74 ~ 81 ~ 88 ~ 95

11. Wendy's pig had a litter of piglets. The five-number summary of the piglets' weights in pounds is given below.
2.1 ~ 2.4 ~ 2.6 ~ 2.9 ~ 3.0
  a. Fifty percent of the piglets weighed less than or equal to ___ pounds.
  b. ___% of the piglets weighed 2.4 pounds or more.
  c. The middle 50% of the piglets weighed between ___ and ___ pounds.

12. A pair of dice were rolled 15 times and the sums were tallied in the table below.

Sum	2	3	4	5	6	7	8	9	10	11	12
Frequency	I	II		I	II	III	III	I	I		I

  a. List the outcomes of the fifteen rolls.
  b. What is the five-number summary of the data?
  c. What is the IQR of the data?
  d. What percent of the sums were between the first and third quartile (Q1 and Q3)?

Use the given information to complete each ordered data set.

13. ___, 4, 7, 9, 10, ___, ___
Five-number-summary = 2 ~ 4 ~ 9 ~ 11 ~ 15

14. ___, ___, 10, 11, 14, ___, ___, 15, 17
Mode = 15   Q1 = 9   Range = 11

15. ___, 14, 18, 20, 23, 23, ___, 30
Range = 16   IQR = 10

16. 68, 74, ___, 81, 82, 82, ___, 95, ___
Q1 = 75   Range = 31   Mean = 83

REVIEW

Identify whether each graph has a good line of best fit drawn. Explain your reasoning.

17.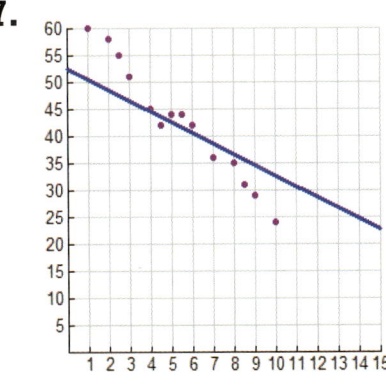

18.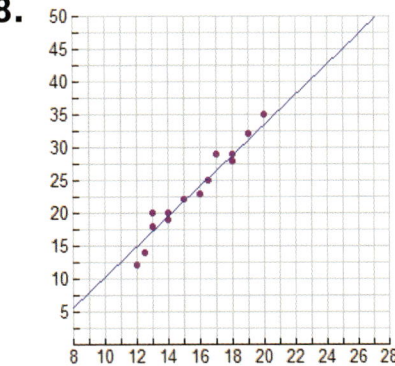

Lesson 5.3 ~ Five-Number Summaries of Data **195**

Use the scatter plot and line of best fit to approximate each missing value.

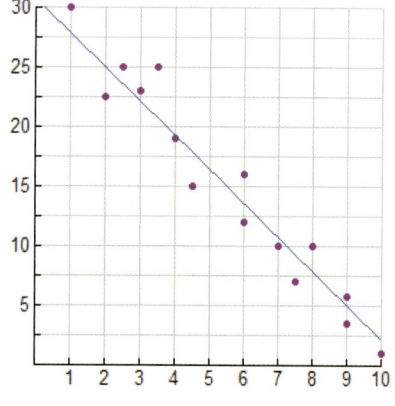

19. When $x = 8$, $y =$ _____

20. When $y = 25$, $x =$ _____

21. When $y = 5$, $x =$ _____

22. Sketch an example of a scatter plot with a positive correlation. Draw a line of best fit. Describe how you determined the placement of your line of best fit.

23. Jackie cooks for a summer camp. For the first week she bought 14 bags of dried fruit and 9 bags of rice for $70.15. For the second week of camp she bought 20 bags of dried fruit and 18 bags of rice for $112.30. How much does a single bag of dried fruit cost? Show all work necessary to justify your answer.

Tic-Tac-Toe ~ Correlation Coefficients

Research correlation coefficients. Make a booklet describing what you've learned. Be sure to address the following questions:
- What is a correlation coefficient and how does it relate to scatter plots?
- What is an "r-value"?
- How can graphing calculators help you find the correlation coefficient?
- How can you tell if data has a "strong" positive correlation?
- How can you tell if data has a "strong" negative correlation?
- What will happen to the r-value if there is no correlation?

Tic-Tac-Toe ~ Test A Theory

A conjecture is a statement thought to be true but not proven. Think of two conjectures about students in school that states a relationship between two variables. For example, "As students get older, the amount of homework they are assigned increases."

Conduct a survey of at least 30 students at school. Collect the data necessary to test the conjectures. Make a scatter plot showing each set of data and analyze the results. Describe the correlations of the data sets. Were the conjectures correct? Was there anything surprising about the results of the survey?

Lesson 5.3 ~ Five-Number Summaries of Data

Q-POINTS AND LINES OF BEST FIT

LESSON 5.4

 Write equations for a line of best fit based on Q-points.

Lines of best fit approximate the pattern of a data set. They are helpful for making predictions. A common problem with lines of best fit is that two students may draw their lines differently. This could greatly affect predictions made from those lines.

To eliminate this problem, statisticians have developed ways to determine a line of best fit. One method for finding a line of best fit is to find the **Q-points** of the data set. The Q-points (quartile points) of a data set are the points created by the intersection of the quartiles for the *x*- and *y*-values. Follow the steps below to find the line of best fit for a given data set.

Using Q-Points to Find a Line of Best Fit

1. Draw a scatter plot of the data set.
2. Find the five-number summary for the *x*-values. Draw vertical lines on the scatter plot at the Q1 and Q3 values.
3. Find the five-number summary for the *y*-values. Draw horizontal lines on the scatter plot at the Q1 and Q3 values.
4. Find the four intersection points created by the lines in **parts 2 and 3**. Write the ordered pairs of the two Q-points that follow the pattern of the data.
5. Write a slope-intercept form equation for the line that passes through the two Q-points.

EXAMPLE 1

Mrs. Harvey's PE class collected data about the effect of workout time on weight loss. Each student recorded how many hours they spent working out for a month. They also recorded how much weight they lost. Use the Q-points of the data to find a line of best fit.

Hours Spent Working Out	3	5	8	8	9	14	17	20	24	24	28
Number of Pounds Lost	2	3	3	6	5	6	7	9	8	10	10

SOLUTION

Draw a scatter plot of the data.

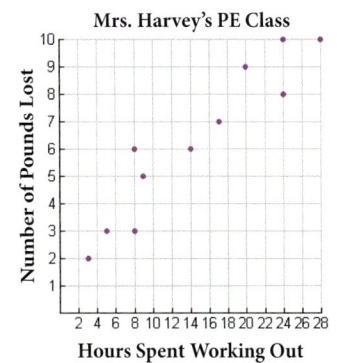

Lesson 5.4 ~ Q-Points and Lines of Best Fit 197

EXAMPLE 1
SOLUTION
(CONTINUED)

Find the five-number summary of the hours spent working out (x-values).
$$3, 5, 8, 8, 9, 14, 17, 20, 24, 24, 28$$
$$3 \sim 8 \sim 14 \sim 24 \sim 28$$

Draw vertical lines at Q1 = 8 and Q3 = 24.

Find the five-number summary of the y-values. Be sure to list the y-coordinates in order.
$$2, 3, 3, 5, 6, 6, 7, 8, 9, 10, 10$$
$$2 \sim 3 \sim 6 \sim 9 \sim 10$$

Draw horizontal lines at Q1 = 3 and Q3 = 9.

From the four intersection points created by the lines, the points (8, 3) and (24, 9) follow the pattern of the data the best. These are the two Q-points that lie on the line of best fit.

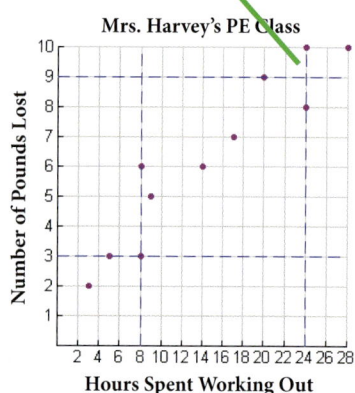

A Q-point may not actually be a point in the data set.

Find the equation of the line through the two points. Write the equation in slope-intercept form: $y = mx + b$.

$$m = \text{slope} = \frac{y_2 - y_1}{x_2 - x_1} = \frac{9 - 3}{24 - 8} = \frac{6}{16} = \frac{3}{8} \quad \rightarrow \quad y = \frac{3}{8}x + b$$

Substitute one of the points (8, 3) into the equation to find the y-intercept (b):
$$3 = \frac{3}{8}(8) + b$$
$$3 = 3 + b$$
$$0 = b$$

The equation for the line of best fit based on Q-points is $y = \frac{3}{8}x + 0$ or $y = \frac{3}{8}x$.

The process of finding a line of best fit by Q-points has several steps. Be sure to carefully go through the process to arrive at an accurate equation.

EXPLORE!   THE WAVE

Your team has just scored the winning touchdown in the football game. The crowd goes wild and starts to do 'the wave' to celebrate the victory. How long will it take the wave to travel from one end of the stadium and back?

Step 1: Have four students in the class line up to do the wave. A student at one end of the line throws his or her hands up and brings them back down. When the first student's hands are up, the next student does the same, creating a 'wave' down the line. When the 'wave' reaches the end of the line, it should reverse back to the start of the line. Record the time it took to complete the wave to the nearest second.

EXPLORE! (CONTINUED)

Step 2: Copy the table below. Continue to collect data for various numbers of students in the wave line. Each trial should have a different number of students. Collect eleven trials.

Number of Students in Wave	4										
Time (seconds)											

Step 3: Let x represent the number of students in the wave. Let y represent the time in seconds. Plot the ordered pairs from the table on graph paper.

Step 4: Find the five-number summary for the x-values and the y-values. Be sure each set of values is written in order before finding the five-number summary.

Step 5: Draw vertical lines on the graph from **Step 4** at the quartiles (Q1 and Q3) for the x-values. Draw horizontal lines on the graph at the quartiles for the y-values.

Step 6: Write the ordered pairs for the four Q-points created by the intersecting lines. Circle the two ordered pairs that best follow the pattern of the data.

Step 7: Find the slope between the two points chosen in **Step 6**. Show how to calculate the slope.

Step 8: Substitute the slope and the coordinates of one of the two Q-points in the appropriate places in the equation: $y = mx + b$. Solve the equation for the y-intercept, b.

Step 9: Write the final equation for the line in slope-intercept form. Check with a classmate. Do you have the same equation? If not, work together to determine the correct equation.

Finally, use the equation to make a prediction…

Step 10: A stadium has 6 sections that stretch across one side. Each section has 30 seats per row. How many people could be seated in each row across the whole side of the stadium?

Step 11: Substitute the answer from **Step 10** into the **Step 9** equation. Since this number represents the number of people in the 'wave', substitute it for x. Evaluate the expression to find the time, in seconds, to complete the 'wave' with this number of people. Is this the exact answer? Explain your reasoning.

Lesson 5.4 ~ Q-Points and Lines of Best Fit

EXAMPLE 2

Use the scatter plot to answer the questions. The vertical and horizontal lines based on the quartiles have already been drawn.

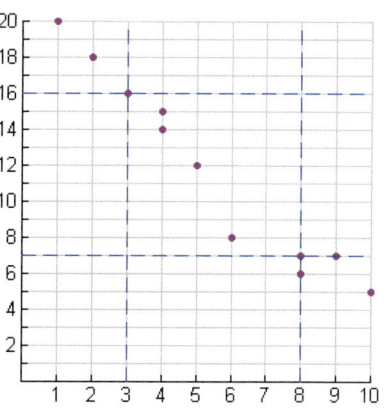

a. Write the ordered pairs for the four Q-points created by the intersecting lines.
b. Does the graph show a positive or negative correlation? Which two Q-points follow the direction of the data best?
c. What is the slope of the line between those Q-points?
d. Write an equation in slope-intercept form for the equation of the line through the Q-points.

SOLUTIONS

a. The ordered pairs are (8, 16), (8, 7), (3, 7) and (3, 16).

b. Because the *y*-values decrease as *x* increases, this scatter plot shows a negative correlation. The two Q-points that follow the data best are (3, 16) and (8, 7).

c. The slope of the line through these Q-points is:

$$Slope = \frac{y_2 - y_1}{x_2 - x_1} = \frac{7 - 16}{8 - 3} = \frac{-9}{5} = -1.8$$

d. Slope-Intercept form: $y = mx + b$
    Substitute the slope from **part c**. $y = -1.8x + b$
    Substitute one of the points, (8, 7). $7 = -1.8(8) + b$
    Solve for *b*. $7 = -14.4 + b$
    $21.4 = b$
    Write the final equation. $y = -1.8x + 21.4$

EXERCISES

1. Why do you think mathematicians developed the Q-points method for finding lines of best fit? What makes this method better for finding a line of best fit than using the method learned in **Lesson 5.2**?

2. What does the "Q" in Q-points represent?

3. Susan drew a scatter plot of a data set. She drew vertical and horizontal lines at Q1 and Q3 of the *x*- and *y*-values. How should she decide which of the points to use for her two Q-points?

Write the ordered pairs for the four Q-points for each graph. Mark the two ordered pairs which best follow the data.

4.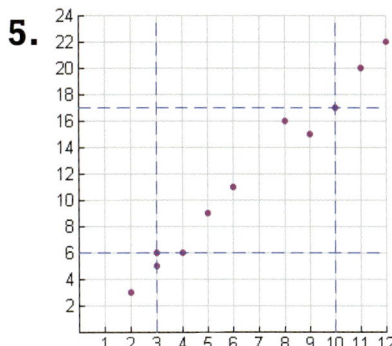

5.

6.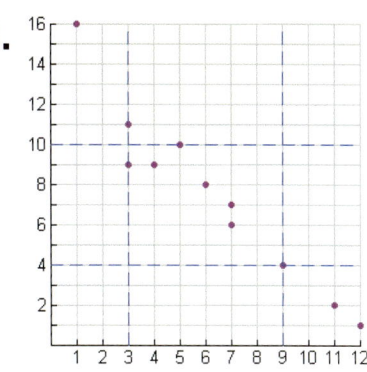

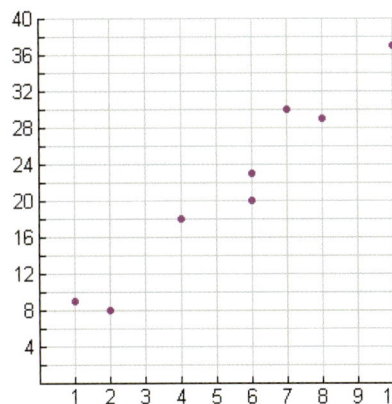

7. Use the scatter plot at left.
   a. Write the ordered pairs for the points that are shown in the scatter plot.
   b. Find the five-number summary of the *x*-coordinates.
   c. Find the five-number summary of the *y*-coordinates.
   d. Copy the scatter plot and draw the horizontal and vertical lines from the quartiles in **parts b and c**.
   e. Write the ordered pairs for the two Q-points that follow the data best.
   f. Were these points in the list of ordered pairs from **part a**? Explain.

Write the equation for the line through the given Q-points in slope-intercept form.

8. (5, 21) and (11, 3)

9. (4, 8) and (12, 20)

10. The following table shows the cost of U.S. postage stamps since 1975. Follow the steps below to write an equation for the line of best fit based on Q-points.

Year	1975	1978	1981	1985	1988	1991	1995	1999	2001	2006	2011
Cost (cents)	13	15	18	22	25	29	32	33	34	39	44

Source: Time Almanac

   a. Let *x* represent the years and let *y* represent the cost of a stamp in cents. Draw a scatter plot of the data.
   b. Find the five-number summary for the *x*- and *y*-values.
   c. Draw the horizontal and vertical lines from your quartiles into the scatter plot.
   d. Write the ordered pairs for the two Q-points that follow the pattern of the data best.
   e. Write an equation for the line through the Q-points from **part d** in slope-intercept form. Round numbers to the nearest tenth.

11. Find the equation of the line of best fit based on Q-points for the scatter plot shown in **Exercise 6**.

Lesson 5.4 ~ Q-Points and Lines of Best Fit **201**

12. The following table shows the number of missing assignments students had in English class and their overall grade as a percent. Use the data to find the equation for the line of best fit based on Q-points.

Number of Missing Assignments (x)	0	0	0	1	1	2	3	3	4	5	6	6	7	9	10
Overall Percent (y)	98	82	95	95	90	92	78	88	82	85	80	68	75	65	50

REVIEW

13. Louise was doing a study to see if there is a relationship between a person's age and the number of times they eat per day. She collected data from fourteen people and made the following scatter plot.

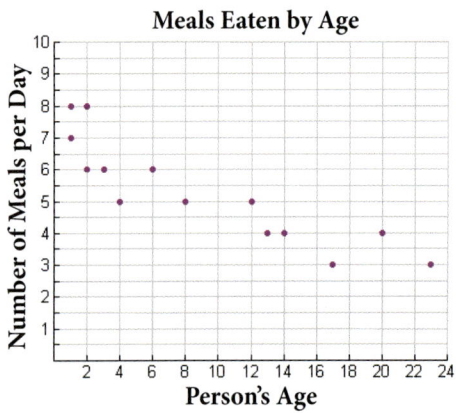

a. Does there appear to be a relationship between a person's age and the number of times they eat per day? Describe the correlation of the data.

b. Louise only collected data for people 23 years old or younger. Do you think this scatter plot could be used to predict how many meals per day a 50-year-old person would eat? Explain your reasoning.

14. Draw a scatter plot for the following set of data and describe the correlation.

Average SAT Scores (Critical Reading Section)

Year (x)	1988	1998	1980	2009	2002	1990	1976	2000	2011	1984	1994	1996	2004
Average Score (y)	505	505	502	501	504	500	509	505	497	504	499	505	508

Source: U.S. Department of Education, National Center for Education Statistics

15. Using the equation $y = 2x - 9$, find the value of x when $y = 25$.

16. Using the equation $y = -\frac{3}{4}x + 1$, find the value of x when $y = 5\frac{1}{2}$.

17. Using the equation $y = 0.6x + 7.4$, find the value of x when $y = 8.24$.

PREDICTING WITH BEST FIT EQUATIONS

LESSON 5.5

 Use equations based on data to make predictions and judge the reasonableness of predictions.

Marlon sells ice cream cones in the park during the summer. He collected information about the high temperature on different days and the number of ice cream cones he sold on each day. He used this information to create a scatter plot and found the equation for a line of best fit. He determined that the number of ice cream cones he sells (N) is related to the high temperature in degrees Fahrenheit (t) using the equation:

$$N = 50 + 0.8t$$

The equation for his line of best fit can be used to make predictions about future events. One summer day the temperature is 95° F. How many ice cream cones could Marlon expect to sell?

Substitute 95 for t in the equation.
Simplify.

$N = 50 + 0.8t$
$N = 50 + 0.8(95)$
$N = 50 + 76 = 126$

Marlon can expect to sell approximately 126 ice cream cones when the high temperature is 95° F.

EXAMPLE 1 Use the equation and given value to solve for the value of the missing variable.
a. $y = 2x - 8$
  $x = 11, y = $ ___
b. $T = 40 + 1.6d$
  $T = 65, d = $ ___

SOLUTIONS

a. Write the original equation.
Substitute 11 for x in the equation.
Simplify.

$y = 2x - 8$
$y = 2(11) - 8$
$y = 22 - 8$
$y = 14$

b. Write the original equation.
Substitute 65 for T in the equation.
Subtract 40 from both sides of the equation.
Divide both sides by 1.6.

$T = 40 + 1.6d$
$65 = 40 + 1.6d$
$\underline{-40 \quad -40}$
$\dfrac{25}{1.6} = \dfrac{1.6d}{1.6}$
$15.625 = d$

EXAMPLE 2

A grocery store manager tracked the cost of a gallon of milk for the last several months of 2011. The line of best fit models the price, P, based on the number of months (m) since December 2011: $P = 3.60 + 0.04m$.

a. What value of m would represent January 2013?
b. Use the equation $P = 3.60 + 0.04m$ to predict the price of a gallon of milk at this store in January 2013.
c. How many months after December 2011 will a gallon of milk first cost more than $5.00 at this store?

SOLUTIONS

a. January 2013 is a full year (12 months) plus a month after December 2011, so $m = 13$.

b. January 2013 means $m = 13$.
Substitute 13 into the equation for m.

$P = 3.60 + 0.04m$
$P = 3.60 + 0.04(13)$

Evaluate the equation.

$P = 3.60 + 0.52$
$P = 4.12$

A gallon of milk will cost $4.12 in January 2013.

c. The price of milk is $5.00.
Substitute 5.00 for P and solve for m.
Subtract 3.60 from both sides.
Divide both sides by 0.04.

$P = 3.60 + 0.04m$
$5.00 = 3.60 + 0.04m$
$-3.60 \quad -3.60$
$\dfrac{1.40}{0.04} = \dfrac{0.04m}{0.04}$
$35 = m$

After 35 months, milk will cost $5.00 at this grocery store.

Now you have two methods for using two-variable data to make predictions.

Extend the Line of Best Fit

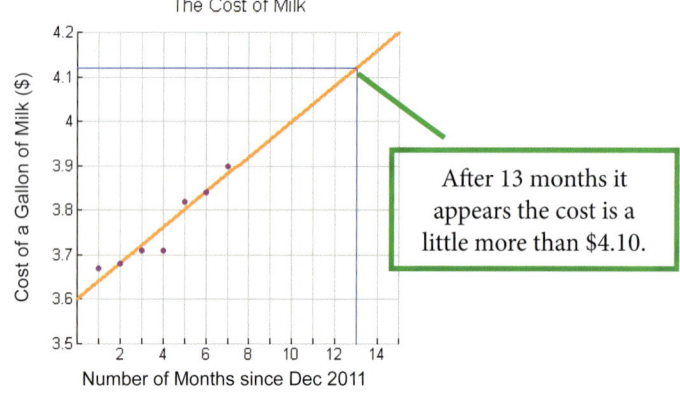

After 13 months it appears the cost is a little more than $4.10.

Use the Equation for a Line of Best Fit

$P = 3.60 + 0.04m$
$P = 3.60 + 0.04(13)$
$P = 3.60 + 0.52$
$P = 4.12$

EXAMPLE 3 Cybil works in the maternity section of the hospital as a nurse. She determined that a baby's weight in pounds (P) can be modeled by $P = 7.2 + 0.6w$, where w is the baby's age in weeks. Does this model seem reasonable for a one-year old baby?

SOLUTION A one-year old baby is 52 weeks old.
Evaluate the equation for $w = 52$.

$P = 7.2 + 0.6w$
$P = 7.2 + 0.6(52)$
$P = 7.2 + 31.2$
$P = 38.4$ pounds

A weight of 38.4 pounds is far too heavy for a 1-year old baby. This model may only be reasonable for the first few months after a baby's birth.

EXERCISES

Use the equation and given value to solve for the value of the missing variable.

1. $y = 3x + 10$; $x = 12$, $y =$ ___

2. $y = -4x + 42$; $y = 10$, $x =$ ___

3. $P = 24 + 0.8n$; $P = 37.60$, $n =$ ___

4. $T = 40 - 1.6d$; $d = 15$, $T =$ ___

5. Mrs. Gonzalez noticed that class sizes at her school have been slowly declining over the years. When she started teaching in 1980, there was an average of 29 students in her classes. Now she only has an average of 24 students per class. Using a line of best fit, she found that the number of students per class (y) can be modeled by $y = 29 - 0.2x$, where x represents the number of years since 1980.
   a. What value of x would represent the year 2020?
   b. Use the equation to predict the number of students per class in the year 2020.
   c. In what year will the student population reach 18 students per class?
   d. Does this equation accurately predict the class sizes at this school in the year 2050? Explain your reasoning.

6. Mike and Aurora organize whitewater rafting trips with their friends. The more people they can get to join them on a trip, the less it costs per person. They determine that the cost per person (C) can be modeled by $C = 35 - 0.75p$, where p represents the number of people rafting.
   a. If 15 people go on the rafting trip, how much will it cost per person?
   b. How many people need to go on the rafting trip for the cost to decrease to $20 per person?
   c. If 50 people go on the rafting trip, how much will it cost? Is it reasonable to use this equation for a 50-person rafting trip? Explain your reasoning.

7. The graph at right shows the winning times (rounded to the nearest tenth) for the Men's 100-Meter Dash in the Summer Olympics. The horizontal and vertical lines representing the quartiles have already been drawn.
   a. Describe the correlation of the data.
   b. Name the two Q-points that follow the direction of the data.
   c. The equation of the line of best fit is approximately $y = -0.0083x + 26.53$. Use the equation to predict the winning 100-Meter Dash time in the year 2052.
   d. Use the equation to predict when the winning time will decrease to 8 seconds. Is this reasonable? Explain your reasoning.

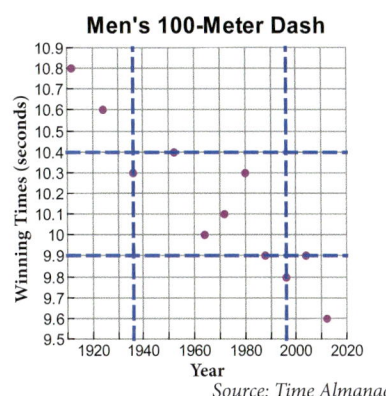

Men's 100-Meter Dash

Source: Time Almanac

8. Ali hikes on the weekends. She started with small hikes and has gradually increased her distance. The first hike she went on was 1.2 miles. The second was 1.5 miles. Use the information in the table to complete the following problems.

Number of Hikes	1	2	3	4	5	6	7	8	9
Distance of Hike (miles)	1.2	1.5	1.6	1.9	2.3	2.3	2.7	2.8	3

a. Let x represent the number of hikes Ali has been on. Let y represent the distance of her hike. Make a scatter plot of the data.
b. Find the five-number summary of the x-values. Draw vertical lines on the graph for the quartile values.
c. Find the five-number summary of the y-values. Draw horizontal lines on the graph for the quartile values.
d. What are the coordinates of the two Q-points that follow the direction of the data?
e. Write an equation for the line through these points.
f. Use the equation to predict how far Ali will travel on her twentieth hike.
g. Predict how many hikes it will take for Ali to reach a 10-mile hike.
h. Will this equation apply for Ali's one-hundredth hike? Explain your reasoning.

REVIEW

9. Use the following ordered pairs to do the following:

x	2	4	4	7	8	10	12
y	20	17	16	11	12	9	4

a. Draw a scatter plot of the data set.
b. Find the four Q-points of the data set. Circle the two Q-points that best follow the pattern of the data.

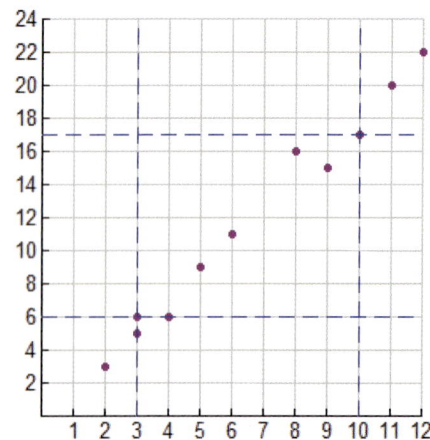

10. Use the scatter plot to answer the following:
    a. What are the four Q-points shown in the scatter plot?
    b. Which two Q-points best follow the pattern of the data?
    c. Write the slope-intercept equation for the line through the Q-points in **part b**.

11. Find the equation of a line with a slope of −3 that goes through the point (4, −7). Show all work necessary to justify your answer.

12. Find the equation of the line that goes through the points (6, 1) and (6, 8). Use words and/or numbers to show how you determined your answer.

USING DATA AND GRAPHS TO PERSUADE

LESSON 5.6

 Recognize and eliminate bias in surveys and data displays.

Data can be extremely powerful in making decisions and predictions. However, it is important to know that data can also be used to persuade and that sometimes data displays can be misleading. Oftentimes it is consumers who are the targets of misleading data.

A common way of collecting data is using a written or verbal survey. The survey is used to collect information about a specific **population**, the group of people or objects being studied. Instead of studying the entire population, a **sample** might be used. A sample is a part of the population that is used to make conclusions about the entire population. For example, if you wanted information about all the students in your school, you might survey students in a specific class. The class would be a sample of the larger school population. For the results of a survey to be accurate, a **random sample** of the population should be used. A random sample is representative of the population being studied, with each person or object having an equal chance of being in the sample.

EXAMPLE 1

For each situation, which option gives the best random sample? Explain your reasoning.
a. What is teenagers' favorite food?
    Option 1: Ask all your friends to name their favorite food.
    Option 2: Ask students in the hallway during passing time.
b. How many books do students typically read per year?
    Option 1: Ask students in the cafeteria during lunch time.
    Option 2: Ask students in the library during lunch time.

SOLUTIONS

a. Option 2. Asking only friends does not give other teenagers the opportunity to be in the sample. Asking students in the hallway would give the best random sample.

b. Option 1. Students in the library at lunch time are more likely to read a lot of books. This does not give a very random sample. Surveying students in the cafeteria is more likely to give a random sample of students' reading habits.

When conducting a survey or analyzing the results of another survey it is important to keep these questions in mind:

Who conducts the survey or data collection can make a big difference in how people respond. Are people going to feel comfortable being truthful to the surveyor?

What questions are asked can greatly affect the results of a survey. Are the questions asked going to lead people to a certain answer?

When a survey is conducted can affect the randomness of the sample. Is each person in the studied population given an equal chance to be included?

Where the survey is given also affects the randomness of the sample. Is the chosen location likely to make certain groups of people more or less likely to be represented?

How the survey is given should be considered. Is it a verbal or written survey? Is responding optional? Optional surveys tend to get responses from people with strong opinions.

EXPLORE! ELIMINATING BIAS

Each survey situation has some sort of **bias**. Bias is a systematic error that contributes to the inaccuracy of a sample.

A. A restaurant owner asks customers to rate their dining experience on a scale of 1-10 as they leave the restaurant.

B. A worker surveys people outside the monkey exhibit: "What is your favorite animal at the zoo?"

C. A pollster questions customers as they leave a local grocery store on a Saturday: "How much do you spend weekly on groceries?"

D. An ice cream company surveys customers, asking: "What is your favorite type of ice cream: Chocolate, Vanilla or Caramel Fudge Swirl?"

E. A restaurant leaves a comment card at each table where customers can choose to rate their dining experience on a scale of 1-10.

WHO? WHAT? WHEN? WHERE? HOW?

Step 1: Identify which type of bias (who, what, when, where, how) each situation could create. Explain your reasoning.

Step 2: Describe how each survey situation could be improved to eliminate the bias.

Another way data can be misleading is in how it is displayed. Choosing certain types of graphs over others, as well as changing the scale of axes, can create very different impressions of a data set.

EXAMPLE 2

Students at Humphrey Middle School collected data on the type of music students would prefer for the next school dance. The bar graph at right shows the results.
a. What is misleading about the graph?
b. Draw a new bar graph which represents the data better.
c. Why might someone use the original graph to represent the data?

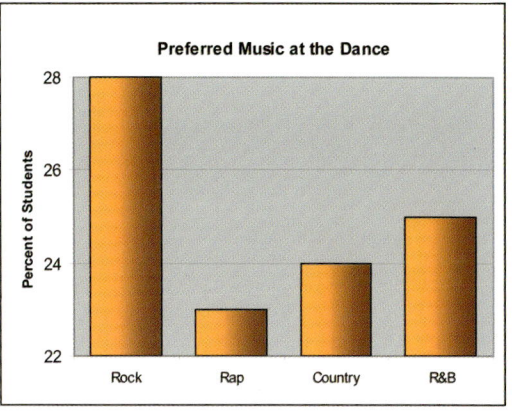

SOLUTIONS

a. The graph does not start at 0%. This makes it appear that Rock had a large majority of the votes. There is really only a five percent difference between Rock and Rap.

b. Starting the graph at 0% would better represent how close the vote was.

c. If someone really wanted to convince people that Rock was a far more popular choice among students, they might use the original graph.

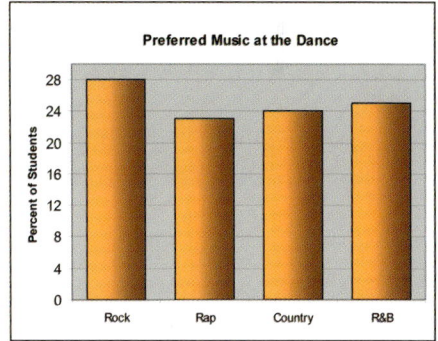

EXERCISES

1. What is the difference between a population and a sample?

2. Why is it important to have a random sample when doing a statistical study?

3. Brad writes for the school newspaper. He surveyed a group of his friends using the survey questions at the right. He called them late the night before his article was due to the newspaper. Some of his friends did not answer the phone.
He drew the following conclusions for his article:
   • Students at Happy Rock Middle School list math as their favorite class.
   • They have an average of 5 A's in their classes.
   • Students typically stay up until about 11 pm.

   > **Students at Happy Rock Middle School**
   > 1. What is your favorite class? (Choose one)
   >   Science ___ Math ___ Music ___
   > 2. In how many classes do you have an A?
   > 3. How late do you stay up on school nights?

   a. Who did Brad survey to get his data? How might this create bias?
   b. What is wrong with the conclusion Brad made from question #1 on his survey?
   c. Is Brad's data about the number of A's that students are earning accurate? What bias might have been created when Brad asked his friends this question?
   d. Brad made his phone calls late at night. How might this have affected the results of his survey?
   e. If Brad did his survey again, what recommendations would improve the accuracy of the results?

Describe the possible bias in each survey situation. Explain how to modify each situation to eliminate the bias.

4. An employee asks customers entering a pizza restaurant about their "favorite type of food".

5. The president of a company asks employees if they are "happy with their salary".

6. Respondents answer a telephone survey about whether they plan to vote Republican or Democrat for the next Presidential election.

7. A girl asks her friends how much allowance they receive per month to determine a typical allowance amount.

8. A hair salon placed a "1-800" number on its receipts so customers can call and give feedback on their haircuts.

9. A teacher asks the first one hundred students who arrive at school on Monday what their GPA is in order to determine the typical GPA of students at her school.

10. The line plot at right shows the gas prices at one gas station from 2009 to 2013.
    a. Draw a new line plot which exaggerates the increase in gas prices since 2009.
    b. Which graph (the original or the graph from **part a**) would the station owner use to bring in more customers? Explain your reasoning.
    c. Which graph would a car dealer use to convince customers to buy a hybrid vehicle? Explain your reasoning.

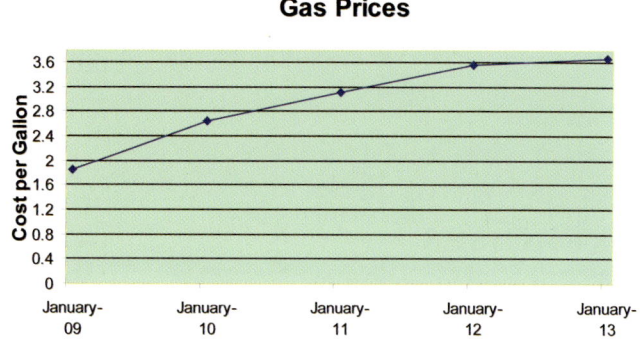

11. Tyrell wants to conduct a survey to determine what type of music his classmates want to hear at an upcoming school dance. Describe a method he could use for surveying students that would give him unbiased data. Explain why this method would be unbiased.

12. Kate lives in a college town. She is curious about what percent of the state cheers for the college in her town and what percent cheers for their rival in another town in the state. She decides to stand outside of the local grocery store and ask the first 100 people that come out to gather data. Is her survey biased or not? Explain your reasoning. If it is biased, explain how she could eliminate the bias.

REVIEW

Use the equation and given value to solve for the value of the missing variable.

13. $y = 58 - 5x$
    $x = 7, y = \_\_\_$

14. $y = 9x + 5$
    $y = 68, x = \_\_\_$

15. $y = 21 - 3x$
    $y = 45, x = \_\_\_$

16. Margaret analyzed her employees' pay using a line of best fit. She found that the hourly salary, S, of her employees can be modeled by $S = 8.15 + 0.45n$, where n represents the number of years in the company.
    a. What is the approximate hourly salary of an employee who has been with the company for 8 years?
    b. About how many years will someone need to work for the company in order to make $14 per hour?

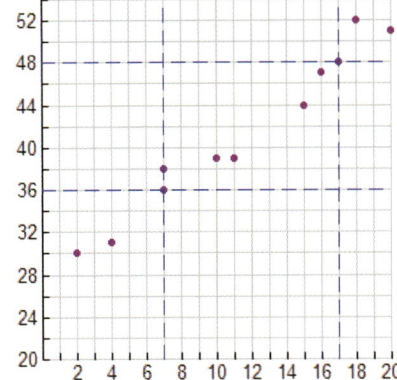

17. Use the scatter plot at left to complete the following. The horizontal and vertical lines representing the quartiles have been drawn in.
    a. Describe the correlation of the data.
    b. Name the two Q-points that follow the direction of the data.
    c. Write a slope-intercept equation for the line through the Q-points.

TIC-TAC-TOE ~ GOOD SURVEY/BAD SURVEY

Lesson 5.6 explained random samples and bias. Use that information to create a good survey. The survey should include at least five questions that are worded appropriately to eliminate bias. Explain how to conduct the survey (the "who", "what", "when", "where", "how" from **Lesson 5.6**). The survey does not have to actually be conducted. Just explain how it should be done and the questions to ask.

Next, take the good survey and turn it into a bad survey. The content of the questions should stay the same. Change how the questions are worded. Also, explain how to conduct this survey. The bad survey should show an understanding of the various forms of bias and a lack of random sampling. You do not actually have to conduct the survey; just explain how it would be done.

TIC-TAC-TOE ~ Q-POINTS POSTER

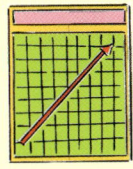

Make a poster that models the process for finding a line of best fit based on Q-points. Use an example to show all the steps involved and include helpful teaching tips. On a separate sheet of paper, create at least two example data sets (with different amounts of points) that students could use to practice. Include a key for these problems.

BIVARIATE DATA AND FREQUENCY TABLES

LESSON 5.7

 Describe associations between two sets of data using relative and conditional frequencies.

When finding the heights of basketball players or listing scores on a test, you are working with single-variable data, also called **univariate data**. The purpose of univariate data is to describe one set of data. Two-variable data, also called **bivariate data**, looks at the relationship between two different sets of data. The purpose of bivariate data is to explain relationships or causes. For example, is there a relationship between the amount of absences a student has and the grade a student earns in class? Is there a relationship between people who own dogs and the number of walks they take?

The amount of absences a student has and the grade a student earns in class are examples of numerical data. To see this relationship, create a scatter plot with x-values representing the number of absences and y-values representing the grade earned in class as a percent. Then, determine if there is a correlation between the data.

Absences x	Grade y
0	95
0	86
0	91
1	85
1	90
1	100
1	88
2	87
3	86
3	90
3	70
7	75
9	72
10	65
20	50

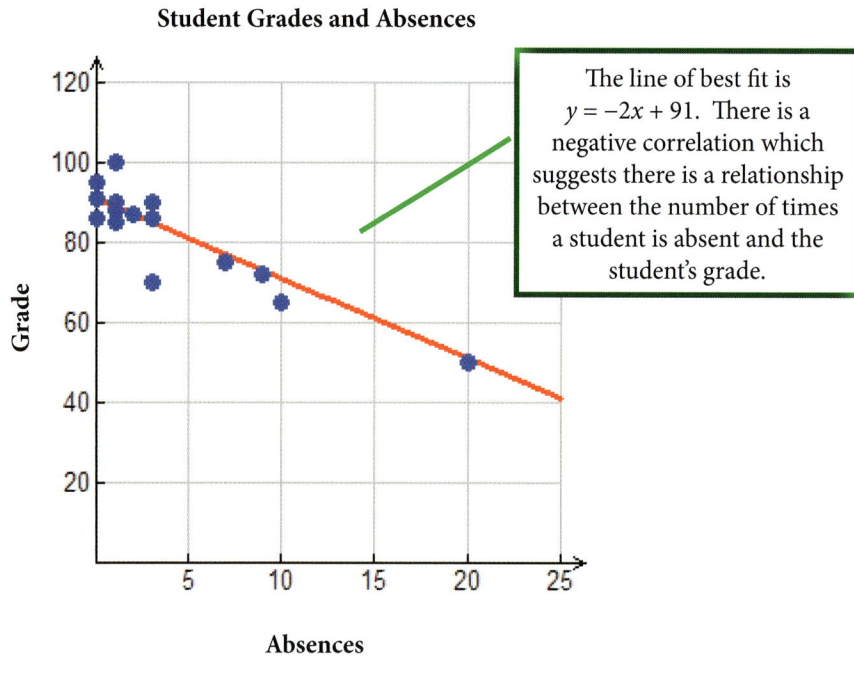

The line of best fit is $y = -2x + 91$. There is a negative correlation which suggests there is a relationship between the number of times a student is absent and the student's grade.

Whether or not a person owns a dog or takes a walk are examples of categorical data. **Categorical data** is data collected in the form of words and frequencies. Frequency is the number of times an item occurs in a data set.

An example of a category is "dog owner." The number of times someone answers he or she is a dog owner would be the frequency of dog owners in the sample. A frequency table shows the number of times each category is represented. When looking at bivariate categorical data, you can use a **two-way frequency table** to see if there is a relationship between two types of categorical data. Scatter plots are not helpful because there are no ordered pairs to plot.

212   Lesson 5.7 ~ Bivariate Data and Frequency Tables

A survey was conducted asking 80 people if they own a dog and if they have taken a walk for exercise within the past two days. The results are below.

- 35 people owned a dog and 45 people did not own dogs.
- Of the 35 people who owned dogs, 15 had walked for exercise in the last two days.
- Of the 45 people who did not own dogs, 25 had walked for exercise in the last two days.

This information can be organized in a two-way frequency table. Determine the number of dog owners who had walked and not walked as well as the number of non-dog owners who had walked and not walked.

	Walk Yes	Walk No
Dog Owners Yes	15	20
Dog Owners No	25	20

Two-way Frequencies
15 dog owners walked.
20 dog owners did not walk.

25 non-dog owners walked.
20 non-dog owners did not walk.

The sum of the numbers in the table equals the number of people surveyed.
15 + 20 + 25 + 20 = 80

EXAMPLE 1

Marsha wanted to investigate a relationship between getting a flu shot and getting the flu. She asked fifty people if they had gotten a flu shot and whether or not they had been sick with the flu.

Her data showed that 40 people had gotten a flu shot and 10 had not.
Of the 40 with a flu shot, 2 had been sick with the flu.
Of the 10 without a flu shot, 7 had been sick with the flu.

Construct a two-way frequency table showing this information.

SOLUTION

Determine the categories.   "Flu Shot" and "Sick with Flu"

Construct a two-way frequency table.

	Sick with Flu Yes	Sick with Flu No
Flu Shot Yes		
Flu Shot No		

Determine the frequencies in each pair of categories.

	Sick with Flu Yes	Sick with Flu No
Flu Shot Yes	2	38
Flu Shot No	7	3

40 − 2 = 38
10 − 7 = 3

Consider the two-way frequency table showing dog owners and whether or not the dog owner walked for exercise in the last two days.

	Walk	
Dog Owner	Yes	No
Yes	15	20
No	25	20

To determine if there is any relationship between dog owners and whether or not a person walked for exercise in the last two days, you can use relative frequencies. A **relative frequency** is the ratio of the observed frequency to the total number of frequencies in the experiment or survey. In the case of the dog owners and walking for exercise, 80 people were surveyed. To find the relative frequencies, find the ratio of each two-way category to 80.

RELATIVE FREQUENCIES

1. Find the total number in the sample.
2. Calculate the ratio of each frequency to the total number sampled.

	Walk	
Dog Owner	Yes	No
Yes	$\frac{15}{80} = \frac{3}{16} \approx 0.19$	$\frac{20}{80} = \frac{1}{4} = 0.25$
No	$\frac{25}{80} = \frac{5}{16} \approx 0.31$	$\frac{20}{80} = \frac{1}{4} = 0.25$

The relative frequencies show:

	Walk	
Dog Owner	Yes	No
Yes	≈ 19% The probability a person owns a dog and has walked is about 19%.	25% The probability a person owns a dog and has not walked is 25%.
No	≈ 31% The probability a person does not own a dog and has walked is about 31%.	25% The probability a person does not own a dog and has not walked is 25%.

These percents are close to one another so there is not a clear relationship between the data. You can also look at conditional frequencies. A **conditional frequency** looks at the ratio within one category. For example, to look at whether a dog owner is more likely to have walked for exercise in the past week, look at the ratio of each frequency for dog owner and compare it to the total dog owners surveyed.

CONDITIONAL FREQUENCIES

1. Find the total number sampled in one category.
2. Calculate the ratio of each frequency in that category to the total sampled in that category.

	Walk	
Dog Owner	**Yes**	**No**
Yes	15	20
No	25	20

Total dog owners surveyed: 15 + 20 = 35

- The conditional frequency for a dog owner who has walked is $\frac{15}{35} = \frac{3}{7} \approx 0.43$

- The conditional frequency for a dog owner who has not walked is $\frac{20}{35} = \frac{4}{7} \approx 0.57$

This means there is a 43% chance a dog owner has walked in the past two days and a 57% chance a dog owner has not walked in the past two days. There is a slightly higher chance if a person is a dog owner that they have not walked in the past two days.

EXAMPLE 2

a. Find the relative frequencies for the two-way table in Example 1 showing the possible relationship between those who had a flu shot and those who were sick with the flu.
b. Explain one observation from the relative frequencies.
c. What is the conditional frequency that someone with a flu shot will not get sick?

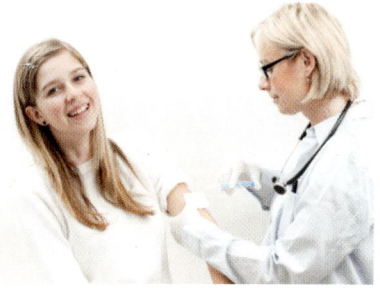

SOLUTIONS

a.
	Sick with Flu	
Flu Shot	**Yes**	**No**
Yes	2	38
No	7	3

Determine the number of people in the survey   2 + 38 + 7 + 3 = 50

Find the ratio of each frequency to the total number surveyed.

	Sick with Flu	
Flu Shot	**Yes**	**No**
Yes	$\frac{2}{50} = \frac{1}{25} = 0.04$	$\frac{38}{50} = \frac{19}{25} = 0.76$
No	$\frac{7}{50} = 0.14$	$\frac{3}{50} = 0.06$

b. One relative frequency shows 76% of those surveyed had a flu shot and did not get sick. It is likely a person asked will have had a flu shot and not gotten sick.

c. Find the number of people who had a flu shot.   2 + 38 = 40
Find the ratio of the frequency to 40. $\frac{38}{40} = \frac{19}{20} = 0.95$

According to this data, there is a 95% chance you will not get the flu if you have a flu shot.

EXERCISES

1. Alex wrote the following information about the two-way frequency table. Which statement is incorrect? Find the statement and correct it.

- 70 people who chose hot dogs used a bun.
- 15 people who chose hot dogs did not use a bun.
- 5 people who did not choose hot dogs used a bun.
- 30 people who did not choose hot dogs did not use a bun.
- There were 100 people surveyed at a picnic about whether they chose to eat a hot dog and whether they chose to eat a bun.

	Bun	
Hot Dog	Yes	No
Yes	70	15
No	5	30

For Exercises 2-3:
   a. **Find the total number of people surveyed.**
   b. **Write each pair of categories with their frequency.**

2.

	Curfew	
Chores	Yes	No
Yes	50	15
No	5	20

3.

	Basketball Player	
Taller than 6 feet	Yes	No
Yes	70	15
No	5	30

4. Two hundred random seniors from a high school were asked if they had passed Algebra II and if they had passed a college readiness exam.
- Of the 120 who had passed Algebra II, 110 had passed the exam.
- Of the 80 who had not passed Algebra II, 30 had passed the exam.

Construct a two-way frequency table showing this information.

5. Ninety people were randomly selected to answer whether or not they liked to fly and whether or not they had a passport.
- Of the 70 people who liked to fly, 60 of them had a passport.
- Fifteen people who did not like to fly had a passport.

Construct a two-way frequency table showing this information.

6. One hundred twenty people were randomly selected to answer whether they preferred country music or rap music and whether they preferred the color green or purple.
- Of the 60 people who preferred country music, 35 preferred the color purple.
- Thirty of the people who preferred rap music preferred the color purple.

Construct a two-way frequency table showing this information.

7. Eighty baseball games for one team were randomly selected from a season of games. Whether the team scored four or more runs and whether the team won was recorded.
- Of the 60 wins the team had scored four or more runs in 45 games.
- The team had scored four or more runs in 8 of the losses.

Construct a two-way frequency table showing this information.

8. Which of the two-way frequency tables in **Exercises 4–7** shows the weakest relationship between the two sets of data? Explain how you know your answer is correct.

9. In a two-way frequency table, what is the difference between a relative frequency and a conditional frequency?

For Exercises 10–11:
  a. Make a two-way relative frequency table. Round to the nearest hundredth.
  b. Explain one observation from the relative frequencies.

10. Home Phone

Cell Phone

	Yes	No
Yes	20	50
No	5	35

11. Play Tennis

Play Golf

	Yes	No
Yes	20	30
No	30	20

12. Write the conditional frequencies for people with cell phones in **Exercise 10**. If a person has a cell phone, are they likely to also have a home phone? Explain your reasoning.

13. Write the conditional frequencies for people who play golf in **Exercise 11**. If a person plays golf, do they likely also play tennis? Explain your reasoning.

For Exercises 14–15, use relative frequencies and/or conditional frequencies to explain whether or not there is a relationship between the bivariate data. Label your work and explain your thinking.

14. Camper

Hiker

	Yes	No
Yes	15	5
No	5	15

15. Play Guitar

Play Piano

	Yes	No
Yes	8	22
No	15	30

REVIEW

Describe the possible bias in each survey situation. Explain how to modify each situation to eliminate the bias.

16. To collect data on Americans' exercise habits, you survey people going in and out of the local fitness center.

17. To collect data on 8th grade students' heights, you measure the height of all the boys in your first period class.

18. Pete researched the cost of flying a group of people from Orlando to Boston for a baseball game. He found that as more people were included in the group, the cost per person went down. The data he collected is in the following table.

Number of People in Group	1	3	5	6	8	9	10	11	12	13	15
Cost per Person (dollars)	350	325	300	295	265	260	255	250	240	230	215

   a. Make a scatter plot of the data. Let x represent the number of people in the group and y represent the cost per person.
   b. Find the quartiles of the x- and y-values. Draw in horizontal and vertical lines on your graph at those values.
   c. Write the ordered pairs for the two Q-points which the data follows.
   d. Find the equation of the line of best fit using the Q-points. Write the values as decimals to the nearest hundredth.
   e. Use your equation to estimate the cost per person if a group had 25 members. Show all work to support your answer.

TIC-TAC-TOE ~ CORRELATIONS IN THE REAL WORLD

Find three examples of data sets with correlations in a magazine or newspaper. The three data sets should include at least one example of a positive correlation and one example of a negative correlation. Cut out or make a copy of each data set. Create a scatter plot, draw a line of best fit and describe the correlation of each data set. Explain why you think each data set has the given correlation.

TIC-TAC-TOE ~ CONDUCT A BIVARIATE SURVEY

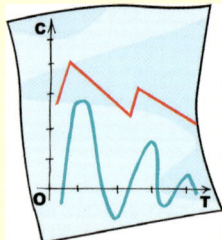

Create a survey with two "Yes or No" questions that you could test to see if there is a relationship. For example, if you want to see if there is a relationship between people who have dogs and the number of walks they take, you might ask the following questions:

Do you own a dog? and *Have you taken a walk in the past two days?*

Survey at least 20 students, asking each student the same two questions, and record the data. Use a two-way frequency table, relative frequencies, and conditional frequencies to analyze your data. Write a 1-page summary of the results from your survey. What conclusions can you draw from your data? Was there any bias in the questions? Did you survey a random sample of students? If you were to do your survey again, would you change anything about the data-collection process?

REVIEW

BLOCK 5

Vocabulary

1st quartile (Q1)	five-number summary	sample
3rd quartile (Q3)	interquartile range (IQR)	scatter plot
bias	line of best fit	single-variable data
bivariate data	median	statistics
categorical data	population	two-variable data
conditional frequency	Q-points	two-way frequency table
correlation	random sample	univariate data
	relative frequency	

Read, create and describe the correlations in scatter plots.
Draw a line of best fit for a set of data.
Use a line of best fit to make predictions.
Find the five-number summary of data sets.
Find the interquartile range (IQR) of data sets.
Write equations for a line of best fit based on Q-points.
Use equations based on data to make predictions and judge the reasonableness of predictions.
Recognize and eliminate bias in surveys and data displays.
Describe associations between two sets of data using relative and conditional frequencies.

Lesson 5.1 ~ Scatter Plots and Correlation

1. Write the ordered pair for each point in the scatter plot at right.

Explain whether each set of data shows positive, negative or no correlation.

2. a person's income and the size of their house

3. the cost of jeans and the number of pairs sold

4. hours spent exercising and the number of calories burned

5. a person's height and the hours they spend exercising

6. Kelby sells hot dogs in the park during the summer. He collected data to show the relationship between the outside temperature and the number of hot dogs he sold.
  a. Make a scatter plot of the data. Use the outside temperatures as the *x*-variable and the number of hot dogs sold as the *y*-variable.
  b. Describe the correlation.

Outside Temperature (Fahrenheit)	Number of Hot Dogs Sold
83°	29
80°	30
95°	13
72°	38
76°	32
88°	24
97°	9
92°	12

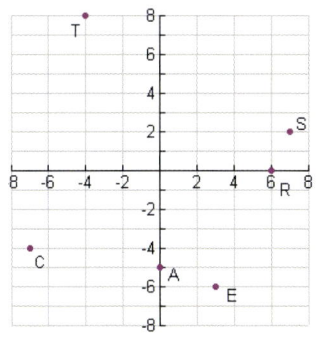

Lesson 5.2 ~ Predicting with Lines of Best Fit

7. What are the characteristics of a good line of best fit?

Identify whether each scatter plot has a good line of best fit drawn. If the line is not a good fit, explain why.

8.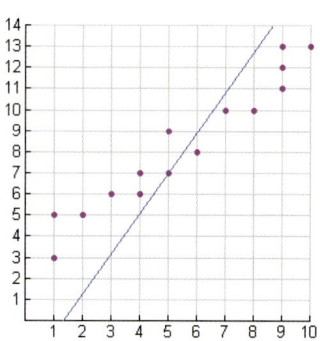

9.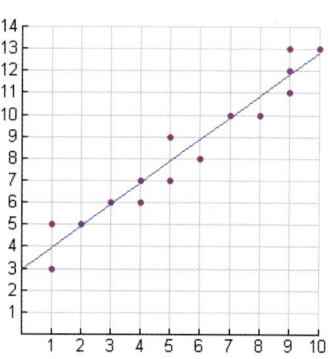

10.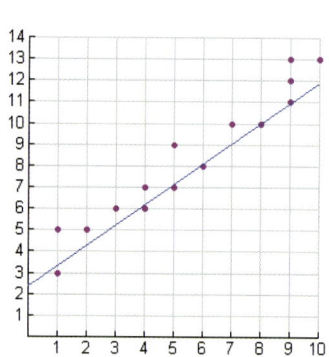

Use the scatter plot and line of best fit to approximate each missing value.

11. When $x = 2020$, $y = $ _____

12. When $y = 30$, $x = $ _____

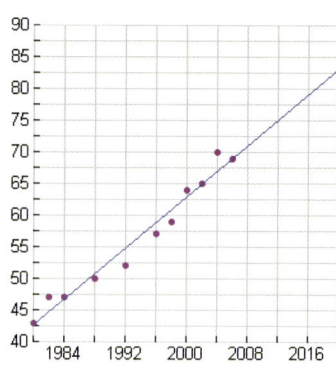

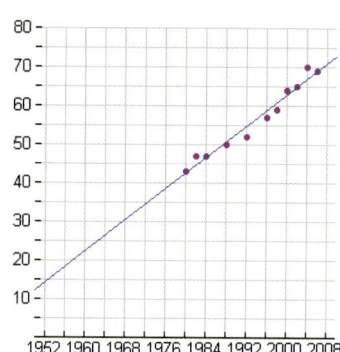

Lesson 5.3 ~ Five-Number Summaries of Data

Find the five-number summary of each data set.

13. 22, 29, 30, 32, 36, 36, 40

14. 4, 10, 12, 17, 19, 20, 28, 30

15. 19, 6, 4, 20, 9, 25, 11, 15, 16

16. 52, 79, 98, 100, 88, 71, 80, 88, 92, 79

Given the five-number summaries, find the interquartile range (IQR) for each data set.

17. 69 ~ 78 ~ 84 ~ 92 ~ 100

18. 20 ~ 23.5 ~ 30 ~ 32 ~ 35

19. For any given data set, what percent of the data is:
  a. Above the 3rd quartile (Q3)?
  b. Between the 1st quartile (Q1) and the 3rd quartile?
  c. Below the median?

20. Use the given information to complete the following ordered data set.

$$2, \_\_, 8, \_\_, 12, 15, 17, \_\_$$
Range = 17   IQR = 10.5   Mean = 10.5

Lesson 5.4 ~ Q-Points and Lines of Best Fit

Write the ordered pairs for the two Q-points that follow the pattern of the data best. Find the equation for the line through those points in slope-intercept form.

21.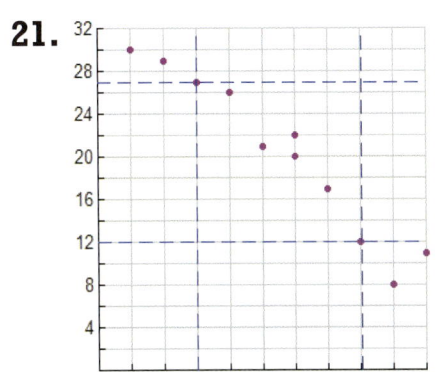

22.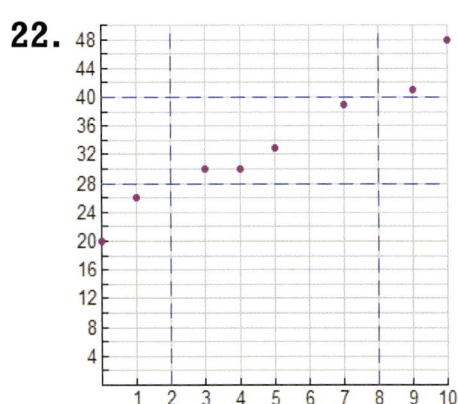

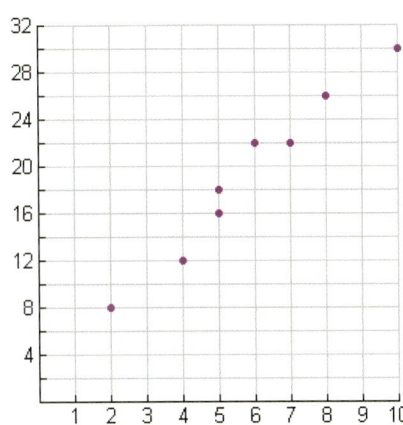

23. Use the scatter plot at left.
   a. Write the ordered pairs for the points that are shown in the scatter plot.
   b. Find the five-number summary of the x-coordinates.
   c. Find the five-number summary of the y-coordinates.
   d. Copy the scatter plot and draw in the horizontal and vertical lines from the quartiles in **parts b and c**.
   e. Write the ordered pairs for the two Q-points that follow the pattern of the data best.

24. Nolan works at a ski resort. He noticed there was a correlation between the number of inches of new snow that day and the number of skiers per hour on a particular ski-run. Use the data to find the equation of a line of best fit based on Q-points. Show all work necessary to justify your answer.

Number of Inches of New Snow (x)	1	4	5	5	6	7	7	8	9	10	10
Number of Skiers per Hour (y)	25	30	32	35	34	38	40	45	51	53	56

Lesson 5.5 ~ Predicting with Best Fit Equations

Use the equation and given value to solve for the value of the missing variable.

25. $y = 48 - 3x$
$x = 9, y =$ _____

26. $y = 7x + 25$
$y = 67, x =$ _____

27. Jessica makes and sells jewelry. She determines her profit, P, from selling the jewelry using the equation $P = 14j - 42$, where j represents the number of pieces of jewelry sold.
   a. Jessica makes and sells 10 pieces of jewelry in one month. How much profit will she make?
   b. How many pieces of jewelry will Jessica need to sell in order to make $1,000?

28. Tanner worked to improve his maximum bench-press lift. Using a scatter plot, he determined that the weight he can lift, L, can be modeled by $L = 185 + 2.4w$, where w represents the number of weeks he has been lifting.
   a. Tanner lifts weights for 8 weeks. How much will he be able to bench-press?
   b. How many weeks will Tanner need to lift to be able to bench-press 221 pounds?
   c. If Tanner lifts weights for 100 weeks, how much will he be able to bench-press according to this model? Is this realistic? Explain your reasoning.

Lesson 5.6 ~ Using Data and Graphs to Persuade

Describe how there could be bias for each survey situation. Explain how each could be changed to eliminate the bias.

29. You ask twenty of your friends what their favorite movie is in order to collect data on students' favorite movies.

30. An employee surveys customers outside of an ice cream shop to determine how many times per month people eat ice cream.

31. A movie critic hands out mail-in questionnaires to people leaving the theater to determine whether people liked the movie they just saw.

32. A student asks other students at school to choose their favorite fruit from bananas, apples and strawberries.

33. The president of a company asks employees if they are "happy with their job" after the employee received a pay raise.

34. The graph at right shows the change in the cost of a digital camera over one year. What is misleading about the graph? Draw a new graph that is not misleading.

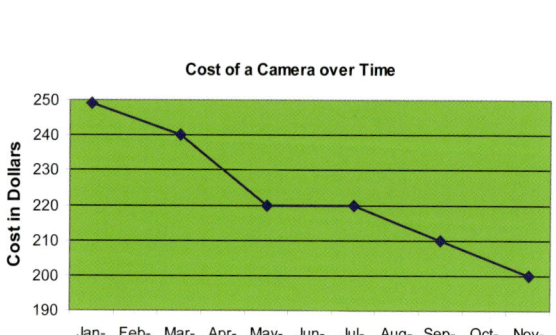

Lesson 5.7 ~ Bivariate Data and Frequency Tables

For Exercises 35-36:
    a. Find the total number of people surveyed.
    b. Write each pair of categories with their frequency.

35. Ride a Skateboard

Ride a Bike	Yes	No
Yes	40	30
No	5	5

36. Own a Smart Phone

Own an MP3 Player	Yes	No
Yes	45	30
No	30	15

37. Sixty students were randomly selected to answer whether or not they were late to first period and whether or not they had ridden the bus to school that morning.
- Of the 30 students who rode the bus, 20 were on time to first period.
- Fifteen students who did not ride the bus were on time to first period.

Construct a two-way frequency table showing the information.

38. One hundred twenty people were randomly selected to answer whether or not they liked salad and whether or not they liked hamburgers.
- Of the 75 people who liked hamburgers, 15 did not like salad.
- Forty people who did not like hamburgers liked salad.

Construct a two-way frequency table showing the information.

39. A survey was conducted asking students whether or not they had purchased lunch or brought lunch to school during the last week. The results are below.

Brought Lunch

Purchased Lunch	Yes	No
Yes	15	40
No	40	5

    a. Make a two-way relative frequency table. Explain one observation from the relative frequencies.
    b. Write the conditional frequencies for people who brought a lunch to school. If a person brought a lunch to school, are they likely to also have purchased a lunch at school that week?

TIC-TAC-TOE ~ GLOSSARY FLASHCARDS

Create flashcards for all of the vocabulary terms in this block. Write the term on one side of the flashcard. The other side of the flashcard should have a definition, a picture, a formula and/or an example.

Quiz yourself on the various vocabulary terms. Create three stacks of flashcards. The first stack should include terms that you know well. The second stack should include terms that you know fairly well, but still need to study. The third stack should include terms that you do not remember very well. Write a short reflection about what terms you still need to study. Turn in the flashcards and reflection.

CAREER FOCUS

BOB
ACTUARY

I am an actuary. An actuary places a value on things that could be lost, broken or destroyed. For instance, suppose it took you a year to save up enough money to purchase an MP3 player. When you buy the MP3 player, the clerk tells you that if it breaks in the first six months the store will replace it. The clerk also tells you that for a few more dollars, the store will replace it if anything goes wrong for two years. The person who calculates how much needs to be paid to guarantee that your MP3 player will be fixed is called an actuary. Actuaries determine how much should be paid for different kinds of insurance to guarantee there will be enough money to replace or fix whatever is wrong.

As an actuary, I use math every day. I use probabilities to calculate things like how many MP3 players will quit working in the first two years. I use division to decide how much each person who wants insurance on their MP3 player should pay for that insurance.

In order to become an actuary, I earned a Bachelor's degree in Mathematics. I also earned a Masters in Actuarial Science. I had to pass 10 exams to achieve the highest actuarial designation offered by the Society of Actuaries. That designation is the Fellow of the Society of Actuaries.

Most actuaries start out making between $44,000 - $61,000 per year. After passing a few of the exams and working for a few years, many actuaries can make over $100,000 per year.

I enjoy my job. I like to provide assurance for people that their valuable items are protected from loss and damage. Insurance helps protect people when unfortunate events occur. As an actuary, I am able to help insurance companies know how much to charge.

ACKNOWLEDGEMENTS

All Photos and Clipart ©2008 Jupiterimages Corporation and Clipart.com with the exception of the cover photo and the following photos:

Linear Equations Page 3
©iStockphoto.com/CandyBox Images

Linear Equations Page 5
©iStockphoto.com/TheCrimsonMonkey

Linear Equations Page 6
©iStockphoto.com/Svemir

Linear Equations Page 7
©iStockphoto.com/Catherine Yeulet

Linear Equations Page 8
©iStockphoto.com/winhorse

Linear Equations Page 9
©iStockphoto.com/Jitalia17

Linear Equations Page 11
©iStockphoto.com/James McQuillan

Linear Equations Page 12
©iStockphoto.com/mangostock

Linear Equations Page 14
©iStockphoto.com/Spiritartist

Linear Equations Page 17
©iStockphoto.com/Lisa F. Young

Linear Equations Page 18
©iStockphoto.com/Rich Seymour

Linear Equations Page 19
©iStockphoto.com/Eric Isselée

Linear Equations Page 19
©iStockphoto.com/londoneye

Linear Equations Page 23
©iStockphoto.com/Jokic

Linear Equations Page 28
©iStockphoto.com/SZE FEI WONG

Linear Equations Page 29
©iStockphoto.com/1 design

Linear Equations Page 32
©iStockphoto.com/kali9

Linear Equations Page 32
©iStockphoto.com/Don Chen Flashon Studios

Linear Equations Page 34
©iStockphoto.com/Adam Turner

Linear Equations Page 34
©iStockphoto.com/Ingvar Bjork

Linear Equations Page 37
©iStockphoto.com/ugar bariskan

Linear Equations Page 37
©iStockphoto.com/Hal Bergman

Linear Equations Page 40
©iStockphoto.com/Christopher Futcher

Linear Equations Page 41
©iStockphoto.com/lisathephotographer

Linear Equations Page 41
©iStockphoto.com/Elliot Westacott

Linear Equations Page 42
©iStockphoto.com/Dennis Owusu-Ansah

Linear Equations Page 42
©iStockphoto.com/sack

Linear Equations Page 46
©iStockphoto.com/Stacey Newman

Linear Equations Page 53
©iStockphoto.com/Sandra Henderson

Linear Equations Page 56
©iStockphoto.com/Thomas Gordon

Linear Equations Page 60
©iStockphoto.com/Lena Sergeeva

Linear Equations Page 62
©iStockphoto.com/PIKSEL

Linear Equations Page 63
©iStockphoto.com/Lara Belova

Linear Equations Page 65
©iStockphoto.com/Stephan Zabel

Linear Equations Page 67
©iStockphoto.com/Bruck Block

Linear Equations Page 71
©iStockphoto.com/svetikd

Linear Equations Page 71
©iStockphoto.com/jeff gynane

Linear Equations Page 75
©iStockphoto.com/Okcaha Tkahyk

Linear Equations Page 76
©iStockphoto.com/kali9

Linear Equations Page 77
©iStockphoto.com/Roberto Manderioli

Linear Equations Page 84
©iStockphoto.com/Suprijono Suharjoto

Linear Equations Page 85
©iStockphoto.com/Josh Campbell

Linear Equations Page 90
©iStockphoto.com/Skip ODonnell

Linear Equations Page 91
©iStockphoto.com/Irina Smirnova

Linear Equations Page 98
©iStockphoto.com/Paul Lampard

Linear Equations Page 104
©iStockphoto.com/Ziva_K

Linear Equations Page 105
©iStockphoto.com/Vasko Miokovic Photography

Linear Equations Page 108
©iStockphoto.com/RiverNorthPhotography

Linear Equations Page 108
©iStockphoto.com/David H. Lewis

Linear Equations Page 112
©iStockphoto.com/Andresr

Linear Equations Page 132
©iStockphoto.com/studiovespa

Linear Equations Page 139
©iStockphoto.com/Evgeniy Ayupov

Linear Equations Page 141
©iStockphoto.com/DElight

Linear Equations Page 143
©iStockphoto.com/Marcin Sadlowski

Linear Equations Page 144
©iStockphoto.com/Nicolas Loran

Linear Equations Page 144
©iStockphoto.com/PIKSEL

Linear Equations Page 145
©iStockphoto.com/Lisa F. Young

Linear Equations Page 148
©iStockphoto.com/Daniel Gale

Linear Equations Page 151
©iStockphoto.com/Forest Woodward

Linear Equations Page 161
©iStockphoto.com/Kristina Hopper

Linear Equations Page 165
©iStockphoto.com/Jurii Konoval

Linear Equations Page 165
©iStockphoto.com/AdShooter

Linear Equations Page 166
©iStockphoto.com/Eliza Snow

Linear Equations Page 178
©iStockphoto.com/4x6

Linear Equations Page 179
©iStockphoto.com/YinYang

Linear Equations Page 187
©iStockphoto.com/blackred

Linear Equations Page 192
©iStockphoto.com/londoneye

Linear Equations Page 195
©iStockphoto.com/Tatiana Belova

Linear Equations Page 200
©iStockphoto.com/paul kline

Linear Equations Page 202
©iStockphoto.com/Catherine Yeulet

Linear Equations Page 203
©iStockphoto.com/Ekash

Linear Equations Page 206
©iStockphoto.com/diego cervo

Linear Equations Page 207
©iStockphoto.com/nisimo

Linear Equations Page 210
©iStockphoto.com/Stephane LARCHER

Linear Equations Page 215
©iStockphoto.com/Wojciech Gajda

Linear Equations Page 216
©iStockphoto.com/Steve McSweeny

Linear Equations Page 216
©iStockphoto.com/YAO MENG PENG

Linear Equations Page 217
©iStockphoto.com/Michael Krinke

Linear Equations Page 217
©iStockphoto.com/Galina Barskaya

Linear Equations Page 221
©iStockphoto.com/Wojciech Gajda

Layout and Design by Judy St. Lawrence

Cover Design by Schuyler St. Lawrence

Glossary Translation by Keyla Santiago and Heather Contreras

CORE FOCUS ON MATH
GLOSSARY ~ GLOSARIO

A

Absolute Value	The distance a number is from 0 on a number line.	**Valor Absoluto**	La distancia de un número desde el 0 en una recta numérica.
Acute Angle	An angle that measures more than 0° but less than 90°.	**Ángulo Agudo**	Un ángulo que mide mas 0° pero menos de 90°.
Adjacent Angles	Two angles that share a ray.	**Ángulos Adyacentes**	Dos ángulos que comparten un rayo.
Algebraic Expression	An expression that contains numbers, operations and variables.	**Expresiones Algebraicas**	Una expresión que contiene números, operaciones y variables.
Alternate Exterior Angles	Two angles that are on the outside of two lines and are on opposite sides of a transversal.	**Ángulos Exteriores Alternos**	Dos ángulos que están afuera de dos rectas y están a lados opuestos de una transversal.
Alternate Interior Angles	Two angles that are on the inside of two lines and are on opposites sides of a transversal.	**Ángulos Interiores Alternos**	Dos ángulos que están adentro de dos rectas y están a lados opuestos de una transversal.
Angle	A figure formed by two rays with a common endpoint.	**Ángulo**	Una figura formada por dos rayos con un punto final en común.

Area	The number of square units needed to cover a surface.	Área	El número de unidades cuadradas necesitadas para cubrir una superficie.
Ascending Order	Numbers arranged from least to greatest.	Progresión Ascendente	Los números ordenados de menor a mayor.
Associative Property	A property that states that numbers in addition or multiplication expressions can be grouped without affecting the value of the expression.	Propiedad Asociativa	Una propiedad que establece que los números en expresiones de suma o de multiplicación pueden ser agrupados sin afectar el valor de la expresión.
Axes	A horizontal and vertical number line on a coordinate plane. 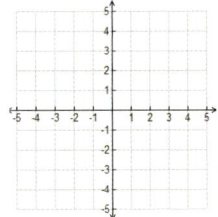	Ejes	Una recta numérica horizontal y vertical en un plano de coordenadas. 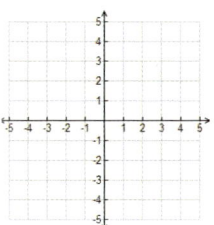
Axis of Symmetry	The line of symmetry on a parabola that goes through the vertex. 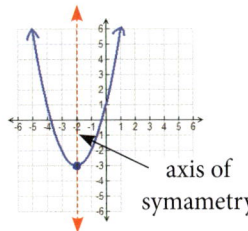 axis of symametry	El Eje De Las Simetría	La linia de simetría de una parábola que pasa por el vértice. 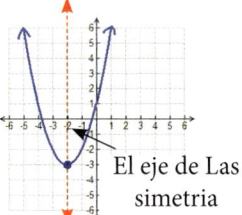 El eje de Las simetria

B

Bar Graph	A graph that uses bars to compare the quantities in a categorical data set. 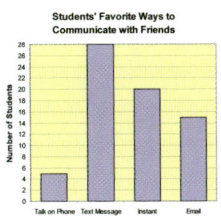	Gráfico de Barras	Una gráfica que utiliza barras para comparar las cantidades en un conjunto de datos categórico. 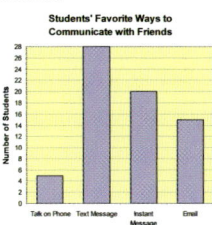
Base (of a power)	The base of the power is the repeated factor. In x^a, x is the base.	Base (de la potencia)	La base de la potenciación es el factor repatidio. En x^a, x es la base.

Glossary ~ Glosario **227**

Base (of a solid)	See Prism, Cylinder, Pyramid and Cone.	Base (de un sólido)	Ver Prisma, Cilindro, Pirámide y Cono.
Base (of a triangle)	Any side of a triangle.	Base (de un triángulo)	Cualquier lado de un triángulo.
Bias	A problem when gathering data that affects the results of the data.	Sesgo	Un problema que ocurre cuando se recogen datos que afectan los resultados de los datos.
Biased Sample	A group from a population that does not accurately represent the entire population.	Muestra Sesgada	Un grupo de una población que no representa con exactitud la población entera.
Binomials	Expressions involving two terms (i.e. $x - 2$).	Binomiales	Expresiones que impliquen dos terminos. (es decir: $x - 2$).
Bivariate Data	Data that describes two variables and looks at the relationship between the two variables.	Datos de dos Variables	Los datos que describen dos variables y analiza la relación entre estas dos variables.
Box-and-Whisker Plot	A diagram used to display the five-number summary of a data set.	Diagrama de Líneas y Bloques	Un diagrama utilizado para mostrar el resumen de cinco números de un conjunto de datos.

C

Categorical Data	Data collected in the form of words.	Datos Categóricos	Datos recopilados en la forma de palabras.
Center of a Circle	The point inside a circle that is the same distance from all points on the circle.	Centro de un Círculo	Un ángulo dentro de un círculo que está a la misma distancia de todos los puntos en el círculo.

Central Angle	An angle in a circle with its vertex at the center of a circle.	Ángulo Central	Un ángulo en un círculo con su vértice en el centro del círculo.
Chord	A line segment with endpoints on a circle.	Cuerda	Un segmento de la recta con puntos finales en el círculo.
Circle	The set of all points that are the same distance from a center point.	Círculo	El conjunto de todos los puntos que están a la misma distancia de un punto central.
Circumference	The distance around a circle.	Circunferencia	La distancia alrededor de un círculo.
Coefficient	The number multiplied by a variable in a term.	Coeficiente	El número multiplicado por una variable en un término.
Commutative Property	A property that states numbers can be added or multiplied in any order.	Propiedad Conmutativa	Una propiedad que establece que los números pueden ser sumados o multiplicados en cualquier orden.
Compatible Numbers	Numbers that are easy to mentally compute; used when estimating products and quotients.	Números Compatibles	Números que son fáciles de calcular mentalmente; utilizado cuando se estiman productos y cocientes.
Complementary Angles	Two angles whose sum is 90°.	Ángulos Complementarios	Dos ángulos cuya suma es de 90°.
Complements	Two probabilities whose sum is 1. Together they make up all the possible outcomes without repeating any outcomes.	Complementos	Dos probabilidades cuya suma es de 1. Juntos crean todos los posibles resultados sin repetir alguno.

Glossary ~ Glosario

Completing the Square	The creation of a perfect square trinomial by adding a constant to an expression in the form $x^2 + bx$.	Terminado el Cuadrado	La creación de un trinomio cuadrado perfecto por adición de una constante a una expresión en la forma $x^2 + bx$.
Complex Fraction	A fraction that contains a fractional expression in its numerator, denominator or both. $$\frac{\frac{3}{4}}{\frac{3}{8}}$$	Fracción Compleja	Una fracción que contiene una expresión fraccionaria en su numerador, el denominador o ambos. $$\frac{\frac{3}{4}}{\frac{3}{8}}$$
Composite Figure	A geometric figure made of two or more geometric shapes.	Figura Compuesta	Una figura geométrica formada por dos o más formas geométricas.
Composite Number	A whole number larger than 1 that has more than two factors.	Número Compuesto	Un número entero mayor que el 1 con más de dos factores.
Composite Solid	A solid made of two or more three-dimensional geometric figures.	Sólido Compuesto	Un sólido formado por dos o más figuras geométricas tridimensionales.
Composition of Transformations	A series of transformations on a point.	Composición de Transformaciones	Una serie de transformaciones en un punto.
Compound Probability	The probability of two or more events occurring.	Compuesto de Probabilidad	La probabilidad de dos o más eventos ocurriendo.
Conditional Frequency	The ratio of the observed frequency to the total number of frequencies in a given category from an experiment or survey.	Frecuencia Condicional	La relación de una frecuencia observada para el número total de frecuencias en una categoría dada del experimento o encuesta.
Cone	A solid formed by one circular base and a vertex.	Cono	Un sólido formado por una base circular y una vértice.
Congruent	Equal in measure.	Congruente	Igual en medida.

Congruent Figures	Two shapes that have the exact same shape and the exact same size. 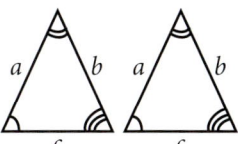	Figuras Congruentes	Dos figuras que tienen exactamente la misma forma y el mismo tamaño. 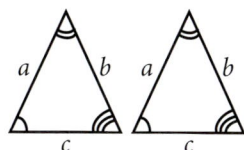
Constant	A term that has no variable.	Constante	Un término que no tiene variable.
Continuous	When a graph can be drawn from beginning to end without any breaks.	Continuo	Cuando una gráfica puede ser dibujada desde principio a fin sin ninguna interrupción.
Conversion	The process of renaming a measurement using different units.	Conversión	El proceso de renombrar una medida utilizando diferentes unidades.
Coordinate Plane	A plane created by two number lines intersecting at a 90° angle.	Plano de Coordenadas	Un plano creado por dos rectas numéricas que se intersecan a un ángulo de 90°. 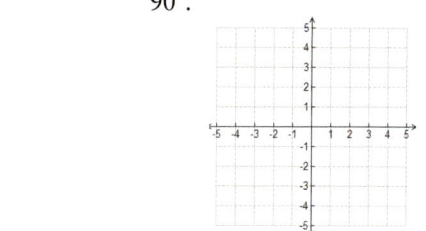
Correlation	The relationship between two variables in a scatter plot.	Correlación	La relación entre dos variables en un gráfico de dispersión.
Corresponding Angles	Two non-adjacent angles that are on the same side of a transversal with one angle inside the two lines and the other on the outside of the two lines. 	Ángulos Correspondientes	Dos ángulos no adyacentes que están en el mismo lado de una transversal con un ángulo adentro de las dos rectas y el otro afuera de las dos rectas.
Corresponding Parts	The angles and sides in similar or congruent figures that match.	Partes Correspondientes	Los ángulos y lados en figuras similares o congruentes que concuerdan.

Glossary ~ Glosario **231**

Cube Root	One of the three equal factors of a number. $3 \cdot 3 \cdot 3 = 27 \quad \sqrt[3]{27} = 3$	Raíz Cúbica	Uno de los tres factores iguales de un número. $3 \cdot 3 \cdot 3 = 27 \quad \sqrt[3]{27} = 3$
Cubed	A term raised to the power of 3.	Cubicado	Un término elevado a la potencia de 3.
Cylinder	A solid formed by two congruent and parallel circular bases.	Cilindro	Un sólido formado por dos bases circulares congruentes y paralelas.

D

Decimal	A number with a digit in the tenths place, hundredths place, etc.	Decimal	Un número con un dígito en las décimas, las centenas, etc.
Degrees	A unit used to measure angles.	Grados	Una unidad utilizada para medir ángulos.
Dependent Events	Two (or more) events such that the outcome of one event affects the outcome of the other event(s).	Eventos Dependiente	Dos (o más) eventos de tal manera que el resultado de un evento afecta el resultado del otro evento (s).
Dependent Variable	The variable in a relationship that depends on the value of the independent variable.	Variable Dependiente	La variable en una relación que depende del valor de la variable independiente.
Descending Order	Numbers arranged from greatest to least.	Progresión Descendente	Los números ordenados de mayor a menor.

Diameter	The distance across a circle through the center. 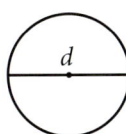	Diámetro	La distancia a través de un círculo por el centro. 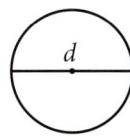
Dilation	A transformation which changes the size of the figure but not the shape. 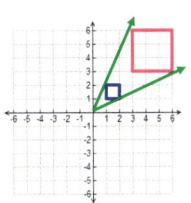	Dilatación	Una transformación que cambia el tamaño de la figura, pero no la forma. 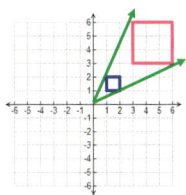
Direct Variation	A linear function that passes through the origin and has the equation $y = mx$.	Variación Directa	Una función lineal que pasa a través del origen y tiene la ecuación $y = mx$.
Discount	The decrease in the price of an item.	Descuento	La disminución de precio en un artículo.
Discrete	When a graph can be represented by a unique set of points rather than a continuous line.	Discreto	Cuando una gráfica puede ser representada por un conjunto de puntos único en vez de una recta continua.
Discriminant	In the quadratic formula, the expression under the radical sign. The discriminant provides information about the number of real roots or solutions of a quadratic equation. $$\frac{-b \pm \sqrt{b^2 - 4ac}}{2a}$$	Discriminante	En la fórmula cuadrática, la expresión bajo el signo radical. El discriminante proporciona información sobre el número o las verdaderas raíces o soluciones de una ecuación cuadrática. $$\frac{-b \pm \sqrt{b^2 - 4ac}}{2a}$$
Distance Formula	A formula used to find the distance between two points on the coordinate plane. $$d = \sqrt{(x_2 - x_1)^2 + (y_2 - y_1)^2}$$	Fórmula de Distancia	Una fórmula utilizada para encontrar la distancia entre dos puntos en un plano de coordenadas. $$d = \sqrt{(x_2 - x_1)^2 + (y_2 - y_1)^2}$$

Distributive Property	A property that can be used to rewrite an expression without parentheses. $$a(b + c) = ab + ac$$	Propiedad Distributiva	Una propiedad que puede ser utilizada para reescribir una expresión sin paréntesis: $$a(b + c) = ab + ac$$
Dividend	The number being divided. $$\mathbf{100} \div 4 = 25$$	Dividendo	El número que es dividido. $$\mathbf{100} \div 4 = 25$$
Divisor	The number used to divide. $$100 \div \mathbf{4} = 25$$	Divisor	El número utilizado para dividir. $$100 \div \mathbf{4} = 25$$
Domain	The set of input values of a function.	Dominio	El conjunto de valores entrados de la función.
Dot Plot	A data display that consists of a number line with dots equally spaced above data values. 	Punto de Gráfico	Una visualización de datos que consiste de una línea numérica con puntos igualmente espaciados sobre valores de datos. 
Double Stem-and-Leaf Plot	A stem-and-leaf plot where one set of data is placed on the right side of the stem and another is placed on the left of the stem. 	Doble Gráfica de Tallo y Hoja	Una gráfica de tallo y hoja donde un conjunto de datos es colocado al lado derecho del tallo y el otro es colocado a la izquierda del tallo. 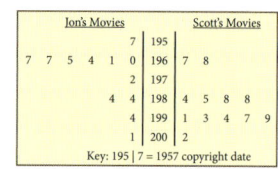

E

Edge	The segment where two faces of a solid meet. 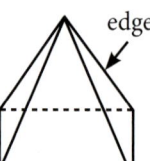	Arista (Borde)	El segmento donde dos caras de un sólido se encuentran. 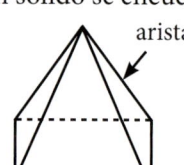

English	Definition	Spanish	Definición
Elimination Method	A method for solving a system of linear equations.	Método de Eliminación	Un método para resolver un sistema de ecuaciones lineales.
Enlargement	A dilation that creates an image larger than its pre-image.	Agrandamiento	Una dilatación que crea una imagen más grande que su pre-imagen.
Equally Likely	Two or more possible outcomes of a given situation that have the same probability.	Igualmente Probables	Dos o más posibles resultados de una situación dada que tienen la misma probabilidad.
Equation	A mathematical sentence that contains an equals sign between 2 expressions.	Ecuación	Una oración matemática que contiene un símbolo de igualdad entre dos expresiones.
Equiangular	A polygon in which all angles are congruent.	Equiángulo	Un polígono en el cual todos los ángulos son congruentes.
Equilateral	A polygon in which all sides are congruent.	Equilátero	Un polígono en el cual todos los lados son congruentes.
Equivalent Decimals	Two or more decimals that represent the same number.	Decimales Equivalentes	Dos o más decimales que representan el mismo número.
Equivalent Expressions	Two or more expressions that represent the same algebraic expression.	Expresiones Equivalentes	Dos o más expresiones que representan la misma expresión algebraica.
Equivalent Fractions	Two or more fractions that represent the same number.	Fracciones Equivalentes	Dos o más fracciones que representan el mismo número.
Evaluate	To find the value of an expression.	Evaluar	Encontrar el valor de una expresión.
Even Distribution	A set of data values that is evenly spread across the range of the data.	Distribución Igualada	Un conjunto de valores de datos que es esparcido de modo uniforme a través del rango de los datos.

Event	A desired outcome or group of outcomes.	Evento	Un resultado o grupo de resultados deseados.
Experimental Probability	The ratio of the number of times an event occurs to the total number of trials.	Probabilidad Experimental	La razón de la cantidad de veces que un suceso ocurre a la cantidad total de intentos.
Exponent	In x^a, a is the exponent. The exponent shows the number of times the factor (x) is repeated.	Exponente	En x^a, a es el exponente. El exponente indica el número de veces que se repite el factor (x).
Exponential Function	A function that can be described by an equation in the form $f(x) = bm^x$.	Función Exponencial	Una función que puede ser descrito por una ecuación en la forma $f(x) = bm^x$.

F

Face	A polygon that is a side or base of a solid.	Cara	Un polígono que es una base de lado de un sólido.
Factors	Whole numbers that can be multiplied together to find a product.	Factores	Números enteros que pueden ser multiplicados entre si para encontrar un producto.
First Quartile (Q1)	The median of the lower half of a data set.	Primer Cuartil (Q1)	Mediana de la parte inferior de un conjunto de datos.
Five-Number Summary	Describes the spread of a data set using the minimum, 1st quartile, median, 3rd quartile and maximum.	Sumario de Cinco Números	Describe la extensión de un conjunto de datos utilizando el mínimo, el primer cuartil, la mediana el tercer cuartil y el máximo.
Formula	An algebraic equation that shows the relationship among specific quantities.	Fórmula	Una ecuación algebraica que enseña la relación entre cantidades específicas.
Fraction	A number that represents a part of a whole number, written as $\frac{numerator}{denominator}$.	Fracción	Un número que representa una parte de un número entero, escrito como $\frac{numerador}{denominador}$.

English	Definition	Spanish	Definición
Frequency	The number of times an item occurs in a data set.	Frecuencia	La cantidad de veces que un artículo ocurre en un conjunto de datos.
Frequency Table	A table which shows how many times a value occurs in a given interval.	Tabla de Frecuencia	Una tabla que enseña cuantas veces un valor ocurre en un intervalo dado.
Function	A relationship between two variables that has one output value for each input value.	Función	Una relación entre dos variables que tiene un valor de salida para cada valor de entrada.

G

English	Definition	Spanish	Definición
General Form	A quadratic function is in general form when written $f(x) = ax^2 + bx + c$ where $a \neq 0$.	Forma General	Una función cuadrática es en forma general cuándo escrito $f(x) = ax^2 + bx + c$ donde $a \neq 0$.
Geometric Probability	Ratios of lengths or areas used to find the likelihood of an event.	Probabilidad Geométrica	Razones de longitudes o áreas utilizadas para encontrar la probabilidad de un suceso.
Geometric Sequence	A list of numbers that begins with a starting value. Each term in the sequence is generated by multiplying the previous term in the sequence by a constant multiplier.	Secuenciación Geométrica	Una lista de números que comienza con un valor inicial. Cada término de la secuencia se genera al multiplicar el término anterior de la secuencia por un multiplicar constante.
Greatest Common Factor (GCF)	The greatest factor that is common to two or more numbers.	Máximo Común Divisor (MCD)	El máximo divisor que le es común a dos o más números.
Grouping Symbols	Symbols such as parentheses or fraction bars that group parts of an expression.	Símbolos de Agrupación	Símbolos como el paréntesis o barras de fracción que agrupan las partes de una expresión.

Glossary ~ Glosario

H

Height of a Triangle	A perpendicular line drawn from the side whose length is the base to the opposite vertex. 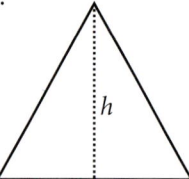	Altura de un Triángulo	Una recta perpendicular dibujada desde el lado cuya longitud es la base al vértice opuesto.
Histogram	A bar graph that displays the frequency of numerical data in equal-sized intervals. 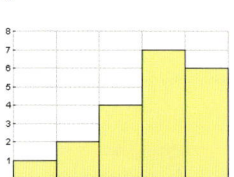	Histograma	Un gráfico de barras que muestra la frecuencia de datos numéricos en intervalos de tamaños iguales. 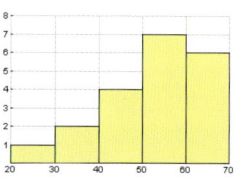
Hypotenuse	The side opposite the right angle in a right triangle. 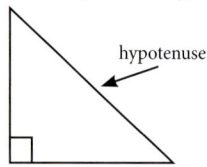	Hipotenusa	El lado opuesto el ángulo recto en un triángulo rectángulo. 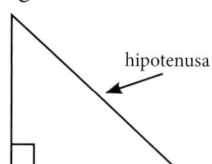

I-J-K

Image	A point or figure which is the result of a transformation or series of transformations.	Imagen	Un punto o figura que es el resultado de una transformación o una serie de transformaciones.
Improper Fraction	A fraction whose numerator is greater than or equal to its denominator.	Fracción Impropia	Una fracción cuyo numerador es mayor o igual a su denominador.
Independent Events	Two (or more) events such that the outcome of one event does not affect the outcome of the other event(s).	Eventos Independientes	Dos (o más) eventos de tal manera que el resultado de un evento no afecta el resultado del otro evento (s).

English	Definition	Spanish	Definición
Independent Variable	The variable representing the input values.	Variable Independiente	La variable que representa los valores entratos.
Inequality	A mathematical sentence using <, >, ≤ or ≥ to compare two quantities.	Desigualdad	Un enunciado matemático usando <, >, ≤ ó ≥ para comparar dos cantidades.
Inference	A logical conclusion based on known information.	Inferencia	Una conclusión lógica basada en la información conocida.
Input-Output Table	A table used to describe a function by listing input values with their output values.	Tabla de Entrada y Salida	Una tabla utilizada para describir una función al enumerar valores de entrada con sus valores de salidas.
Integers	The set of all whole numbers, their opposites, and 0.	Enteros	El conjunto de todos los números enteros, sus opuestos y 0.
Interquartile Range (IQR)	The difference between the 3rd quartile and the 1st quartile in a set of data.	Rango Intercuartil (IQR)	La diferencia entre el tercer cuartil y el primer cuartil en un conjunto de datos.
Inverse Operations	Operations that undo each other.	Operaciones Inversas	Operaciones que se cancelan la una a la otra.
IQR Method	A method for determining outliers using interquartile ranges.	Método IQR	Un método para determinar los datos aberrantes.
Irrational Numbers	A number that cannot be expressed as a fraction of two integers.	Números Irracionales	Un número que no puede ser expresado como una fracción de dos enteros.

Isosceles Trapezoid	A trapezoid that has congruent legs. 	Trapezoide Isósceles	Un trapezoide con catetos congruentes. 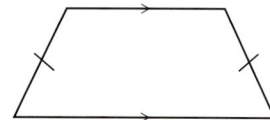
Isosceles Triangle	A triangle that has two or more congruent sides. 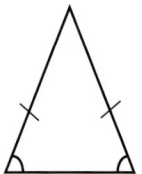	Triángulo Isósceles	Un triángulo que tiene dos o más lados congruentes. 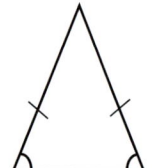

L

Lateral Face	A side of a solid that is not a base.	Cara Lateral	Un lado de un sólido que no sea una base.
Least Common Denominator (LCD)	The least common multiple of two or more denominators.	Mínimo Común Denominador (MCD)	El mínimo común múltiplo de dos o más denominadores.
Least Common Multiple (LCM)	The smallest nonzero multiple that is common to two or more numbers.	Mínimo Común Múltiplo (MCM)	El múltiplo más pequeño que no sea cero que le es común a dos o más números.
Leg	The two sides of a right triangle that form a right angle. 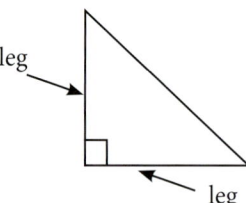	Cateto	Los dos lados de un triángulo rectángulo que forman un ángulo recto. 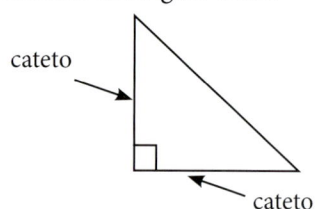
Like Terms	Terms that have the same variable(s).	Términos Semejantes	Términos que tienen el mismo variable(s).

Line of Best Fit	A line which best represents the pattern of a two-variable data set.	Recta de Mejor Ajuste	Una recta que mejor representa el patrón de un conjunto de datos de dos variables.
Linear Equation	An equation whose graph is a line.	Ecuación Lineal	Una ecuación cuya gráfica es una recta.
Linear Function	A function whose graph is a line.	Función Lineal	Una función cuya gráfica es una recta.
Linear Pair	Two adjacent angles whose non-common sides are opposite rays.	Par Lineal	Dos ángulos adyacentes cuyos lados no comunes son rayos opuestos.

M

Mark-up	The increase in the price of an item.	Margen de Beneficio	El aumento de precio en un artículo.
Maximum	The highest point on a curve.	Máximo	El punto más alto en la curva.
Mean	The sum of all values in a data set divided by the number of values.	Media	La suma de todos los valores en un conjunto de datos dividido entre la cantidad de valores.
Mean Absolute Deviation	A statistic that shows the average distance from the mean for all numbers in a data set.	Desviación Media Absoluta	Una estadística que muestra la distancia promedio entre la media de todos los números en una serie de datos.

Glossary ~ Glosario **241**

Measures of Center	Numbers that are used to represent a data set with a single value; the mean, median, and mode are the measures of center.	**Medidas de Centro**	Números que son utilizados para representar un conjunto de datos con un solo valor; la media, la mediana, y la moda son las medidas de centro.
Measures of Variability	Statistics that help determine the spread of numbers in a data set.	**Medidas de Variabilidad**	Las estadísticas que ayudan a determinar la extensión de los números en una serie de datos.
Median	The middle number or the average of the two middle numbers in an ordered data set.	**Mediana**	El número medio o el promedio de los dos números medios en un conjunto de datos ordenados.
Minimum	The lowest point on a curve.	**Mínimo**	El punto más bajo en la curva.
Mixed Number	The sum of a whole number and a fraction less than 1.	**Números Mixtos**	La suma de un número entero y una fracción menor que 1.
Mode	The number(s) or item(s) that occur most often in a data set.	**Moda**	El número(s) o artículo(s) que ocurre con más frecuencia en un conjunto de datos.
Motion Rate	A rate that compares distance to time.	**Índice de Movimiento**	Un índice que compara distancia por tiempo.
Multiple	The product of a number and nonzero whole number.	**Múltiplo**	El producto de un número y un número entero que no sea cero.

N

Negative Number	A number less than 0.	**Número Negativo**	Un número menor que 0.

Net	A two-dimensional pattern that folds to form a solid.	Red	Un patrón bidimensional que se dobla para formar un sólido.
Non-Linear Function	A function whose graph does not form a line.	Ecuación No Lineal	Una ecuación cuya gráfica no forma una recta.
Normal Distribution	A set of data values where the majority of the values are located in the middle of the data set and can be displayed by a bell-shaped curve.	Distribución Normal	Un conjunto de valores de datos donde la mayoría de los valores están localizados en el medio del conjunto de datos y pueden ser mostrados por una curva de forma de campana.
Numerical Data	Data collected in the form of numbers.	Datos Numéricos	Datos recopilados en la forma de números.
Numerical Expressions	An expression consisting of numbers and operations that represents a specific value.	Expresiones Numéricas	Una expresión que consta de números y operaciones que representa un valor específico.

O

Obtuse Angle	An angle that measures more than 90° but less than 180°.	Ángulo Obtuso	Un ángulo que mide más de 90° pero menos de 180°.
Opposites	Numbers that are the same distance from 0 on a number line but are on opposite sides of 0.	Opuestos	Números a la misma distancia del 0 en un recta numérica pero en lados opuestos del 0.
Order of Operations	The rules to follow when evaluating an expression with more than one operation.	Orden de Operaciones	Las reglas a seguir cuando se evalúa una expresión con más de una operación.
Ordered Pair	A pair of numbers used to locate a point on a coordinate plane (x, y).	Par Ordenado	Un par de números utilizados para localizar un punto en un plano de coordenadas (x, y).

Glossary ~ Glosario **243**

Origin	The point where the *x*- and *y*-axis intersect on a coordinate plane (0, 0).	Origen	El punto donde el eje de la *x*- *y* el de la *y*- se cruzan en un plano de coordinadas (0,0).

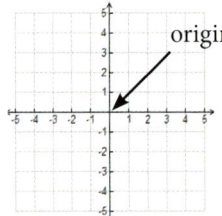
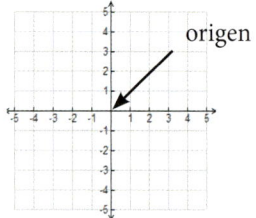

Outcome	One possible result from an experiment or a probability sample space.	Resultado	Un resultado posible de un experimento o un espacio de probabilidad de la muestra.
Outlier	An extreme value that varies greatly from the other values in a data set.	Dato Aberrante	Un valor extremo que varía mucho de los otros valores en un conjunto de datos.

P

Parabola	The graph of a quadratic function.	Parábola	La gráfica de una función cuadratica.

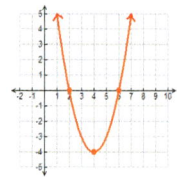

Parallel	Lines in the same plane that never intersect.	Paralela	Rectas en el mismo plano que nunca se intersecan.
Parallel Box-and-Whisker Plot	One box-and-whisker plot placed above another; often used to compare data sets.	Diagrama Paralelo de Líneas y Bloques	Un diagrama de líneas y bloques ubicado sobre otro para comparar conjuntos de datos.

Parallelogram	A quadrilateral with both pairs of opposite sides parallel.	Paralelogramo	Un cuadrilateral con ambos pares de lados opuestos paralelos.
Parent Function	The simplest form of a particular type of function.	Función Principal	La forma más simple de un tipo particular de la función.
Parent Graph	The most basic graph of a function.	Gráfico Matriz	La gráfica más básica de una función.
Percent	A ratio that compares a number to 100.	Por Ciento	Una razón que compara un número con 100.
Percent of Change	The percent a quantity increases or decreases compared to the original amount.	Por Ciento de Cambio	El por ciento que una cantidad aumenta o disminuye comparado a la cantidad original.
Percent of Decrease	The percent of change when the new amount is less than the original amount.	Por Ciento de Disminución	El por ciento de cambio cuando la nueva cantidad es menos que la cantidad original.
Percent of Increase	The percent of change when the new amount is more than the original amount.	Por Ciento de Incremento	El por ciento de cambio cuando la nueva cantidad es más que la cantidad original.
Perfect Cube	A number whose cube root is an integer.	Cubo Perfecto	Un número cuyo raíz cúbica es un número entero.
Perfect Square	A number whose square root is an integer.	Cuadrado Perfecto	Un número cuyo raíz cuadrado es un número entero.
Perfect Square Trinomial	A trinomial that is the square of a binomial.	Trinomio Cuadrado Perfecto	Un trinomio que es el cuadrado de un binomio.
Perimeter	The distance around a figure.	Perímetro	La distancia alrededor de una figura.

Perpendicular	Two lines or segments that form a right angle.	Perpendicular	Dos rectas o segmentos que forman un ángulo recto.
Pi (π)	The ratio of the circumference of a circle to its diameter.	Pi (π)	La razón de la circunferencia de un círculo a su diámetro.
Pictograph	A graph that uses pictures to compare the amounts represented in a categorical data set.	Gráfica Pictórica	Una gráfica que utiliza dibujos para comparar las cantidades representadas en un conjunto de datos categóricos.
Pie Chart	A circle graph that shows information as sectors of a circle.	Gráfico Circular	Enseña la información como sectores de un círculo.
Polygon	A closed figure formed by three or more line segments.	Polígono	Una figura cerrada formada por tres o más segmentos de rectas.
Population	The entire group of people or objects one wants to gather information about.	Población	Todo el grupo de personas o los objetos a los que se quiere obtener información sobre.
Positive Number	A number greater than 0.	Número Positivo	Un número mayor que 0.
Power	An expression such as x^a which consists of two parts, the base (x) and the exponent (a).	Potencia	Una expresión como x^a que consiste de dos partes, la base (x) y el exponente (a).
Pre-image	The original figure prior to a transformation.	Pre-imagen	La figura original antes de una transformación.

English		Spanish	
Prime Factorization	When any composite number is written as the product of all its prime factors.	Factorización Prima	Cuando cualquier número compuesto es escrito como el producto de todos los factores primos.
Prime Number	A whole number larger than 1 that has only two possible factors, 1 and itself.	Número Primo	Un número entero mayor que 1 que tiene solo dos factores posibles, 1 y el mismo.
Prism	A solid formed by polygons with two congruent, parallel bases.	Prisma	Un sólido formado por polígonos con dos bases congruentes y paralelas.
Probability	The measure of how likely it is an event will occur.	Probabilidad	La medida de cuán probable un suceso puede ocurrir.
Product	The answer to a multiplication problem.	Producto	La respuesta a un problema de multiplicación.
Proper Fraction	A fraction with a numerator that is less than the denominator.	Fracción Propia	Una fracción con un numerador que es menos que el denominador.
Proportion	An equation stating two ratios are equivalent.	Proporción	Una ecuación que establece que dos razones son equivalentes.
Protractor	A tool used to measure angles.	Transportador	Una herramienta para medir ángulos.
Pyramid	A solid with a polygonal base and triangular sides that meet at a vertex.	Pirámide	Un sólido con una base poligonal y lados triangulares que se encuentran en un vértice.

Pythagorean Triple	A set of three positive integers (a, b, c) such that $a^2 + b^2 = c^2$.	Triple de Pitágoras	Un conjunto de tres enteros positivos (a, b, c) tal que $a^2 + b^2 = c^2$.

Q

Q-Points	Points that are created by the intersection of the quartiles for the x- and y-values of a two-variable data set.	Puntos Q	Puntos que son creados por la intersección de los cuartiles para los valores de la x- y la y- de un conjunto de datos de dos variables.
Quadrants	Four regions formed by the x and y axes on a coordinate plane. 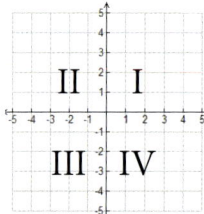	Cuadrantes	Cuatro regiones formadas por el eje-x y el eje-y en un plano de coordinadas. 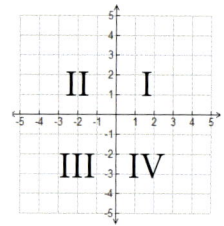
Quadratic Formula	A method which can be used to solve quadratic equations in the form $0 = ax^2 + bx + c$, where $a \neq 0$. $$x = \frac{-b \pm \sqrt{b^2 - 4ac}}{2a}$$	Fórmula Cuadrática	Un métado que puede usarse para resolver ecuaciones cuadraticas en la forma $0 = ax^2 + bx + c$, donde $a \neq 0$. $$x = \frac{-b \pm \sqrt{b^2 - 4ac}}{2a}$$
Quadratic Function	Any function in the family with the parent function of $f(x) = x^2$.	Función Cuadrática	Cualquier otra función en la familia con la función principal de $f(x) = x^2$.
Quadrilateral	A polygon with four sides.	Cuadrilateral	Un polígono con cuatro lados.
Quotient	The answer to a division problem.	Cociente	La solución a un problema de división.

R

Radius	The distance from the center of a circle to any point on the circle.	Radio	La distancia desde el centro de un círculo a cualquier punto en el círculo. 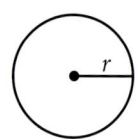

English	Definition	Spanish	Definición
Random Sample	A group from a population created when each member of the population is equally likely to be chosen.	Muestra Aleatoria	Un grupo de una población creada cuando cada miembro de la población tiene la misma probabilidad de ser elegido.
Range (of a data set)	The difference between the maximum and minimum values in a data set.	Rango	La diferencia entre los valores máximo y mínimo en un conjunto de datos.
Range (of a function)	The set of output values of a function.	Rango (de una función)	El conjunto de valores salidos de la función.
Rate	A ratio of two numbers that have different units.	Índice	Una proporción de dos números con diferentes unidades.
Rate Conversion	A process of changing at least one unit of measurement in a rate to a different unit of measurement.	Conversión de Índice	Un proceso de cambiar por lo menos una unidad de medición en un índice a una diferente unidad de medición.
Rate of Change	The change in y-values over the change in x-values on a linear graph.	Índice de Cambio	El cambio en los valores de y sobre el cambio en los valores de x en una gráfica lineal.
Ratio	A comparison of two numbers using division. $$a:b \quad \frac{a}{b} \quad a \text{ to } b$$	Razón	Una comparación de dos números utilizando división. $$a:b \quad \frac{a}{b} \quad a \text{ a } b$$
Rational Number	A number that can be expressed as a fraction of two integers.	Número Racional	Un número que puede ser expresado como una fracción de dos enteros.
Ray	A part of a line that has one endpoint and extends forever in one direction.	Rayo	Una parte de una recta que tiene un punto final y se extiende eternamente en una dirección.
Real Numbers	The set of numbers that includes all rational and irrational numbers.	Números Racionales	El conjunto de números que incluye todos los números racionales e irracionales.

| Reciprocals | Two numbers whose product is 1. | Recíprocos | Dos números cuyo producto es 1. |

| Rectangle | A quadrilateral with four right angles.  | Rectángulo | Un cuadrilátero con cuatro ángulos rectos. |

| Recursive Routine | A routine described by stating the start value and the operation performed to get the following terms. | Rutina Recursiva | Una rutina descrita al exponer el valor del comienzo y la operación realizada para conseguir los términos siguientes. |

| Recursive Sequence | An ordered list of numbers created by a first term and a repeated operation. | Secuencia Recursiva | Una lista de números ordenados creada por un primer término y una operación repetida. |

| Reduction | A dilation that creates an image smaller than its pre-image. | Reducción | Una dilatación que crea una imagen más pequeña que su pre-imagen. |

| Reflection | A transformation in which a mirror image is produced by flipping a figure over a line. 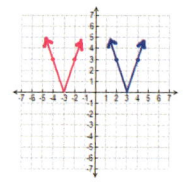 | Reflexión | Una transformación en el que se produce una imagen reflejada volteando una figura sobre una línea. |

| Relative Frequency | The ratio of the observed frequency to the total number of frequencies in an experiment or survey. | Frecuencia Relativa | La proporción de la frecuencia observada para el número total de frecuencias en un experimento o estudio. |

| Remainder | A number that is left over when a division problem is completed. | Remanente | Un número que queda cuando un problema de división se ha completado. |

| Repeating Decimal | A decimal that has one or more digits that repeat forever. | Decimal Repetitivo | Un decimal que tiene uno o más dígitos que se repiten eternamente. |

Representative Sample	A group from a population that accurately represents the entire population.	Muestra Representativa	Un grupo de una población que representa con precisión toda la población.
Rhombus	A quadrilateral with four sides equal in measure. 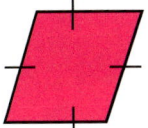	Rombo	Un cuadrilátero con cuatro lados iguales en la medida. 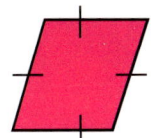
Right Angle	An angle that measures 90°.	Ángulo Recto	Un ángulo que mide 90°.
Roots	The x-intercepts of a quadratic function. 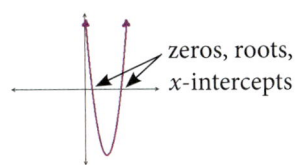 zeros, roots, x-intercepts	Raíces	Las intersecciones-x de una función cuadratica. 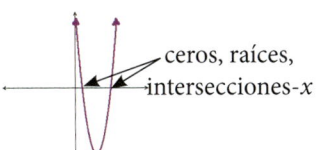 ceros, raíces, intersecciones-x
Rotation	A transformation which turns a point or figure about a fixed point, often the origin. 	Rotación	Una transformación que convierte un punto a una figura sobre un punto fijo. 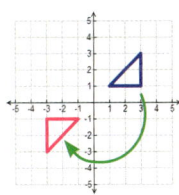

S

Sales Tax	An amount added to the cost of an item. The amount added is a percent of the original amount as determined by a state, county or city.	Impuesto sobre las Ventas	Una cantidad añadida al costo de un artículo. La cantidad añadida es un por ciento de la cantidad original determinado por el estado, condado o ciudad.

Glossary ~ Glosario

Same-Side Interior Angles	Two angles that are on the inside of two lines and are on the same side of a transversal. 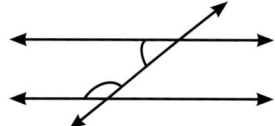	Ángulos Interiores del Mismo Lado	Dos ángulos que están en el interior de dos rectas y están en el mismo lado de una transversal. 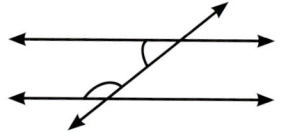
Sample	A group from a population that is used to make conclusions about the entire population.	Muestra	Un grupo de una población que se utiliza para sacar conclusiones sobre toda la población.
Sample Space	The set of all possible outcomes.	Muestra de Espacio	El conjunto de todos los resultados posibles.
Scale	The ratio of a length on a map or model to the actual object.	Escala	La razón de una longitud en un mapa o modelo al objeto verdadero.
Scale Factor	The ratio of corresponding sides in two similar figures.	Factor de Escala	La razón de los lados correspondientes en dos figuras similares.
Scalene Triangle	A triangle that has no congruent sides. 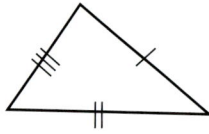	Triángulo Escaleno	Un triángulo sin lados congruentes. 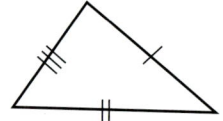
Scatter Plot	A set of ordered pairs graphed on a coordinate plane. 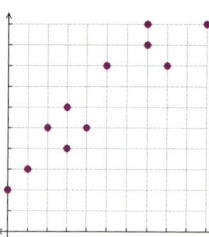	Diagrama de Dispersión	Un conjunto de pares ordenados graficados en un plano de coordenadas. 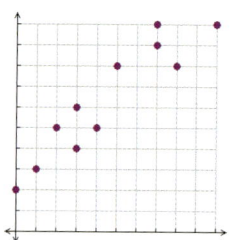
Scientific Notation	Scientific notation is an exponential expression using a power of 10 where $1 \leq N < 10$ and P is an integer. $N \times 10^P$	La Notación Científica	Notación científica es una expresión exponencial con una potencia de 10, donde $1 \leq N < 10$ y P es un número entero. $N \times 10^P$

Sector	A portion of a circle enclosed by two radii.	Sector	Una porción de un circulo encerado por dos radios.
Sequence	An ordered list of numbers.	Sucesión	Una lista de números ordenados.
Similar Figures	Two figures that have the exact same shape, but not necessarily the exact same size.	Figuras Similares	Dos figuras que tienen exactamente la misma forma, pero no necesariamente el mismo tamaño exacto.
Similar Solids	Solids that have the same shape and all corresponding dimensions are proportional.	Sólidos Similares	Sólidos con la misma forma y todas sus dimensiones correspondientes son proporcionales.
Simplest Form	A fraction whose numerator and denominator's only common factor is 1.	Expresión Simple	Una fracción cuyo único factor común del numerador y del denominador es 1.
Simplify an Expression	To rewrite an expression without parentheses and combine all like terms.	Simplificar una Expresión	Reescribir una expresión sin paréntesis y combinar todos los términos iguales.
Simulation	An experiment used to model a situation.	Simulación	Un experimento utilizado para modelar una situación.
Single-Variable Data	A data set with only one type of data.	Datos de una Variable	Un conjunto de datos con tan solo un tipo de datos.
Sketch	To make a figure free hand without the use of measurement tools.	Esbozo	Hacer una figura a mano libre sin utilizar herramientas de medidas.
Skewed Left	A plot or graph with a longer tail on the left-hand side.	Torcido a la Izquierda	Un gráfico con una cola al lado izquierdo.

Skewed Right	A plot or graph with a longer tail on the right-hand side.	Torcido a la Derecha	Un gráfico con una cola al lado derecho.
Slant Height	The height of a lateral face of a pyramid or cone.	Altura Sesgada	La altura de un cara lateral de una pirámide o cono.
Slope	The ratio of the vertical change to the horizontal change in a linear graph.	Pendiente	La razón del cambio vertical al cambio horizontal en una gráfica lineal.
Slope Triangle	A right triangle formed where one leg represents the vertical rise and the other leg is the horizontal run in a linear graph.	Triángulo de Pendiente	Un triángulo rectángulo formado donde una cateto representa el ascenso y la otra es una carrera horizontal en una gráfica lineal.
Slope-Intercept Form	A linear equation written in the form $y = mx + b$.	Forma de las Intersecciones con la Pendiente	Una ecuación lineal escrita en la forma $y = mx + b$.
Solid	A three-dimensional figure that encloses a part of space.	Sólido	Una figura tridimensional que encierra una parte del espacio.
Solution	Any value or values that makes an equation true.	Solución	Cualquier valor o valores que hacen una ecuación verdadera.
Solution of a System of Linear Equations	The ordered pair that satisfies both linear equations in the system.	Solución de un Sistema de Ecuaciones Lineales	El par ordenado que satisface ambas ecuaciones lineales en el sistema.

English		Spanish	
Sphere	A solid formed by a set of points in space that are the same distance from a center point.	Esfera	Un sólido formado por un conjunto de puntos en el espacio que están a la misma distancia de un punto central.
Square	A quadrilateral with four right angles and four congruent sides.	Cuadrado	Un cuadrilátero con cuatro ángulos rectos y cuatro lados congruente.
Square Root	One of the two equal factors of a number. $3 \cdot 3 = 9 \quad 3 = \sqrt{9}$	Raíz Cuadrada	Uno de los dos factores iguales de un número. $3 \cdot 3 = 9 \quad 3 = \sqrt{9}$
Squared	A term raised to the power of 2.	Cuadrado	Un término elevado a la potencia de 2.
Start Value	The output value that is paired with an input value of 0 in an input-output table.	Valor de Comienzo	El valor de salida que es aparejado con un valor de entrada de 0 en una tabla de entradas y salidas.
Statistics	The process of collecting, displaying and analyzing a set of data.	Estadísticas	El proceso de recopilar, exponer y analizar un conjunto de datos.
| Stem-and-Leaf Plot | A plot which uses the digits of the data values to show the shape and distribution of the data set.<br><br>```
 5 | 6
 6 |
 7 | 2 5 9 9
 8 | 0 0 6 8 9
 9 | 2 3 4 8
10 | 0 0
Key: 7 | 5 = 75
``` | Gráfica de Tallo y Hoja | Un diagrama que utiliza los dígitos de los valores de datos para mostrar la forma y la distribución del conjunto de datos.<br><br>```
 5 | 6
 6 |
 7 | 2 5 9 9
 8 | 0 0 6 8 9
 9 | 2 3 4 8
10 | 0 0
Key: 7 | 5 = 75
``` |
| Straight Angle | An angle that measures 180°. 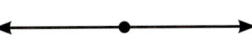 | Ángulo Recto | Un ángulo que mide 180°.  |

English	Definition	Spanish	Definición
Straight Edge	A ruler-like tool with no markings.	Borde Recto	Un gobernante como herramienta sin marcas.
Substitution Method	A method for solving a system of linear equations.	Método de Substitución	Un método para resolver un sistema de ecuaciones lineales.
Supplementary Angles	Two angles whose sum is 180°.	Ángulos Suplementarios	Dos ángulos cuya suma es 180°.
Surface Area	The sum of the areas of all the surfaces on a solid.	Área de la Superficie	La suma de las áreas de todas las superficies en un sólido.
System of Linear Equations	Two or more linear equations.	Sistema de Ecuaciones Lineales	Dos o más ecuaciones lineales.

T

Term	A number or the product of a number and a variable in an algebraic expression; a number in a sequence.	Término	Un número o el producto de un número y una variable en una expresión algebraica; un número en una sucesión.
Terminating Decimal	A decimal that stops.	Decimal Terminado	Un decimal que para.
Theorem	A relationship in mathematics that has been proven.	Teorema	Una relación en las matemáticas que ha sido probada.
Theoretical Probability	The ratio of favorable outcomes to the number of possible outcomes.	Probabilidad Teórica	La proporción de resultados favorables a la cantidad de resultados posibles.
Third Quartile (Q3)	The median of the upper half of a data set.	Tercer Cuartil (Q3)	Mediana de la parte superior de un conjunto de datos.
Tick Marks	Equally divided spaces marked with a small line between every inch or centimeter on a ruler.	Marcas de Graduación	Espacios divididos igualmente marcados con una línea pequeña entre cada pulgada o centímetro en una regla.
Transformation	The movement of a figure on a graph so that it changes size or position.	Transformación	El movimiento de una figura en un gráfico de modo que cambia el tamaño o posición

Translation	A transformation in which a figure is shifted up, down, left or right. 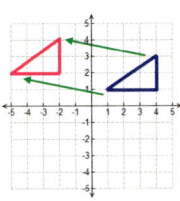	Traducción	Una transformación donde la figura se mudo arriba, abajo, a la izquierda o a la derecha. 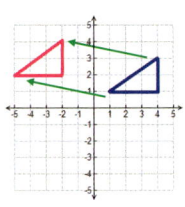
Transversal	A line that intersects two or more lines in the same plane.	Transversal	Una recta que interseca dos o más rectas en el mismo plano.
Trapezoid	A quadrilateral with exactly one pair of parallel sides. 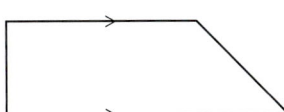	Trapezoide	Un cuadrilateral con exactamente un par de lados paralelos. 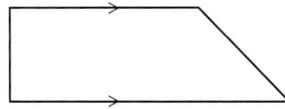
Tree Diagram	A display that organizes information to determine possible outcomes. 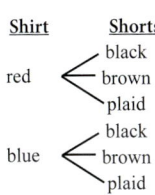	Diagrama de Árbol	Una pantalla que organiza la información para determinar los posibles resulatados. 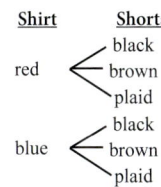
Trial	A single act of performing an experiment.	Prueba	Un solo intento de realizar un experimento.
Trinomial	An expression with three terms (i.e. $x^2 - 3x + 4$).	Trinomio	Una expreción que tiene tres terminos (es decir: $x^2 - 3x + 4$).
Two-Step Equation	An equation that has two different operations.	Ecuación de Dos Pasos	Una ecuación que tiene dos operaciones diferentes.
Two-Variable Data	A data set where two groups of numbers are looked at simultaneously.	Datos de dos Variables	Un conjunto de datos dónde dos grupos de números se observan simultáneamente.

Glossary ~ Glosario **257**

| Two-Way Frequency Table | A table that shows how many times a value occurs for a pair of categorical data. | Tabla de Frecuencia Bidireccional | Una tabla que muestra cuántas veces aparece un valor de un par de datos categóricos. |

	Walk	
Dog Owners	Yes	No
Yes	15	20
No	25	20

	Paseo	
Perro Propietario	Si	No
Si	15	20
No	25	20

U-V-W

| Unit Rate | A rate with a denominator of 1. | Índice de Unidad | Un índice con un denominador de 1. |

| Univariate Data | Data that describes one variable (i.e., scores on a test). | Data Univariados | Datos que describen una variable (es decir: puntajes en una prueba). |

| Variable | A symbol that represents one or more numbers. | Variable | Un símbolo que representa uno o más números. |

| Vertex | The minimum or maximum point on a parabola. | Vértice | El mínimo o máximo punto en una parábola. |

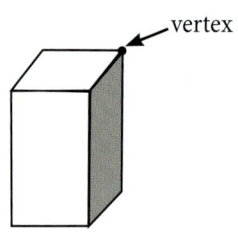
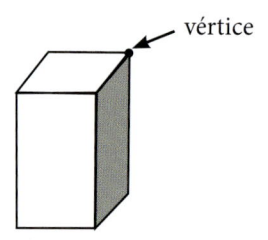

| Vertex of a Solid | The point where three or more edges meet. | Vértice de un Sólido | El punto donde tres o más bordes se encuentran. |

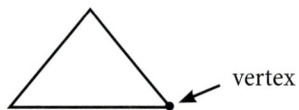

| Vertex of a Triangle | A point where two sides of a triangle meet. | Vértice de un Triángulo | Un punto donde dos lados de un triángulo se encuentran. |

Vertex of an Angle	The common endpoint of the two rays that form an angle. 	Vértice de un Ángulo	El punto final en común de los dos rayos que forma un ángulo. 
Vertex Form	A quadratic function is in vertex form when written $f(x) = a(x - h)^2 + k$ where $a \neq 0$.	Forma De Vértice	Una función cuadrática es en forma general cuándo escrito $f(x) = a(x - h)^2 + k$ donde $a \neq 0$.
Vertical Angles	Non-adjacent angles with a common vertex formed by two intersecting lines. 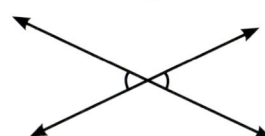	Ángulos Verticales	Ángulos no adyacentes con un vértice en común formado por dos rectas intersecantes. 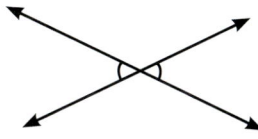
Vertical Line Test	A test used to determine if a graph represents a function by checking to see if a vertical line passes through no more than one point of the graph of a relation.	Examen Vertical De Línia	Un examen para determinar si una gráfica representa una función. Es utilizada para ver si una línia vertical que pasa a través de no más de un punto de la gráfica de una relación.
Volume	The number of cubic units needed to fill a three-dimensional figure.	Volumen	La cantidad de unidades cúbicas necesitadas para llenar un sólido.

X-Y-Z

x-Axis	The horizontal number line on a coordinate plane. 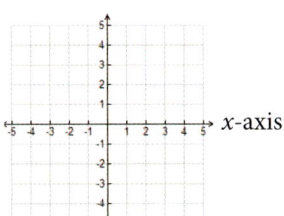	Eje-x, Eje de la x	La recta numérica horizontal en un plano de coordenadas. 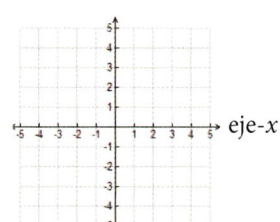

y-Axis	The vertical number line on a coordinate plane. 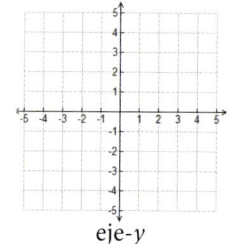 *y*-axis	Eje-*y*, Eje de la *y*	La recta numérica vertical en un plano de coordenadas. eje-*y*
y-Intercept	The point where a graph intersects the *y*-axis. *y*-intercept	Intersección *y*	El punto donde una gráfica interseca el eje-*y*. intersección *y*
Zero Pair	One positive integer chip paired with one negative integer chip. 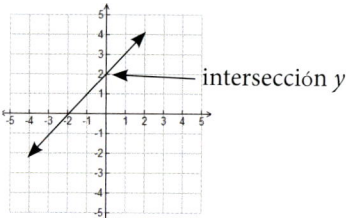 $1 + (-1) = 0$	Par Cero	Un chip entero positivo emparejado con un chip entero negativo. $1 + (-1) = 0$
Zero Product Property	If a product of two factors is equal to zero, then one or both of the factors must be zero.	Propiedad De Producto Cero	Si un producto de dos factores es iqual a cero, uno o ambos de los factores debe ser cero.
Zeros	The *x*-intercepts of a quadratic function. 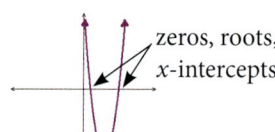 zeros, roots, *x*-intercepts	Ceros	Las intersecciones-*x* de una función cuadratica. 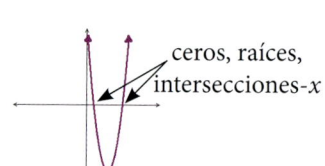 ceros, raíces, intersecciones-*x*

SELECTED ANSWERS

BLOCK 1

Lesson 1.1
1. a) Nathan: multiplication **b)** subtraction **3.** 28 **5.** 17 **7.** 21 **9.** 4 **11.** 8 **13.** 5 **15.** −12 **17. a)** $130 \cdot 5 + 40 \cdot 8$ **b)** $970 **19.** Answers may vary. **21.** $(6 + 3 + 11) \div 4 = 5$ **23.** $-1 \cdot 6 + 8 - 4 \div (2 + 2) = 1$

Lesson 1.2
1. 5 **3.** −6 **5.** 22 **7.** 8 **9.** 30 **11.** 1 **13.** 1
15.

x	6x − 4	Output
−2	6(−2) − 4	−16
0	6(0) − 4	−4
$\frac{1}{2}$	$6(\frac{1}{2}) - 4$	−1
3	6(3) − 4	14
8	6(8) − 4	44

17. true **19.** false **21.** true **23.** He did not multiply 5 and 4; the correct value is 23. **25. a)** 18 units **b)** 26 units **c)** 22 units **27. a)** $12x + 16y$ **b)** $116 **c)** Answers may vary. **29.** −1 **31.** −2 **33.** −53

Lesson 1.3
1. $5x + 5$ **3.** $8m + 10$ **5.** $-3h + 33$ **7.** 735 **9.** 3,584 **11. a)** $7(15 - 0.05)$ **b)** $104.65 **13.** $13x - 8$ **15.** $-x + y$ **17.** $11y + 7x$ **19.** $6x + 1$ **21.** $5x + 10$ **23.** $10x - 8$ **25.** $6x + 18$ **27.** $44x - 77$ **29.** A and C; $3x + 9$ **31. a)** 50 **b)** $\frac{4}{6} = \frac{2}{3}$ **c)** 53 **d)** 32

Lesson 1.4
1. $x = 31$ **3.** $x = 9$ **5.** $x = 61$ **7.** $x = -72$ **9.** $x = 22$ **11.** See student work. $x = 4$ **13.** $x + (-53) = 89$; $x = 142$ **15.** $x - 16 = 102$; $x = 118$ **17.** $\frac{1}{3} \cdot x = 6$; $x = 18$ **19. a)** $7x$ **b)** $7x = 84$; $x = 12$ **21.** $x = -160$ **23.** $x = \frac{7}{4}$ or $1\frac{3}{4}$ **25.** $x = -3.4$ **27.** See student work. **29.** 26 **31.** −6 **33.** $-18x + 2$ **35.** $2x - 1$

Lesson 1.5
1. $x = 4$ **3.** $x = 50$ **5.** $x = 11$ **7.** $x = -60$ **9.** $x = 10$ **11. a)** subtracting instead of adding; 6, $x = 11$ **b)** subtracted from wrong side of equation, $x = 3$ **c)** divided by 8 instead of multiplying; $x = 192$ **13.** $\frac{x}{3} - 6 = -1$; $x = 15$ **15.** $\frac{1}{2}x + 4 = 3$; $x = -2$ **17.** 8.5 hours **19.** 27 fish **21.** $1.08; See student work for explanation. **23.** $2 + 3 \times 5 = 17$ **25.** $-1 \times 2 + 2 = 0$ **27.** $9x - 18$ **29.** $10x$ **31.** $4x + 20$

Lesson 1.6
1. See student work. Subtract x from both sides, then subtract 1 from both sides, then divide both sides by 2; $x = 4$ **3.** $x = 11$ **5.** $x = 1$ **7.** $x = 15$ **9. a)** $12 + 0.25x = 0.75x$ **b)** $x = 24$ **c)** $80 **11.** $x = 8$ **13.** $x = 5$ **15.** $x = 3$ **17.** $x = 4$ **19.** 7 weeks; See student work for explanation. **21.** He forgot to distribute the 3 and the 7; $x = -5$ **23.** true **25.** false

Lesson 1.7
1. Answers may vary. **3.** infinitely many solutions **5.** $x = 5$; one solution **7.** $7 = 7$; infinitely many solutions **9.** $3 = 3$; infinitely many solutions **11.** $-2 = -2$; infinitely many solutions **13.** $x = -4$; one solution **15.** $-0.8 = -0.8$; infinitely many solutions **17.** Logan is correct; the equation simplifies to $-40 = 40$. There are no solutions. **19.** Answers may vary. **21.** $10x + 5$ **23.** $3x - 5$ **25.** $4x + 7y$ **27.** $5x - 35$

Lesson 1.8
1. $x > -3$ **3.** $x \geq 2$
5. $x \geq 4$
7. $x < -4$
9. $x \geq 0$
11. $x \geq 2$
13. $x < -7$
15. $x \leq -2$
17. a) $60x \leq 960$; $x \leq 16$ **b)** $x \leq 14$; See student work to show that 15 boxes would be too many. **19. a)** $0.45x + 5 \leq 12$ **b)** 15 miles **21.** $x = 7$; one solution **23.** $2 = 2$; infinitely many solutions **25.** $x = 3$; one solution

Block 1 Review
1. 18 **3.** 11 **5.** 2 **7.** −11 **9.** 11 **11.** 8.4 **13.** −20 **15.** $\frac{9}{2}$ or $4\frac{1}{2}$
17.

x	$\frac{3x + 2}{4}$	Output
−3	$\frac{3(-3) + 2}{4}$	$-\frac{7}{4}$ or $-1\frac{3}{4}$
0	$\frac{3(0) + 2}{4}$	$\frac{1}{2}$
4	$\frac{3(4) + 2}{4}$	$\frac{7}{2}$ or $3\frac{1}{2}$
10	$\frac{3(10) + 2}{4}$	8

19. $2x - \frac{1}{2}$ **21.** $18x - 60$ **23.** $1.5x + 0.5$ **25.** $3x - 6$ **27.** $5x - 10$ **29.** $10x + 25$ **31.** $5(x - 4) = 5x - 20$ **33.** $x = -21$ **35.** $x = 5.5$ **37.** $x = 22.2$ **39.** $x - 10 = 66$; $x = 76$ **41.** $x = 10$ **43.** $x = 36$ **45.** $x = 5.5$ **47. a)** $72 **b)** $30 + 14m = S$ **c)** $m = 10$ **49.** $x = -2$ **51.** $x = 11$ **53.** $x = -1$ **55.** 6 weeks; See student work for explanation. **57.** Mark is correct. He can divide both sides by −5 first or distribute. Both paths will arrive at the correct solution. **59.** $-15 = -15$; infinitely many solutions **61.** $-1.5 = 5$; no solutions **63.** $10 = 10$; infinitely many solutions **65.** $x = 3.5$; one solution **67.** disagree; the two expressions are not equal **69.** $x > 1$
71. $x > 1$ **73.** $x \leq -4$

Selected Answers **261**

BLOCK 2

Lesson 2.1

1. 3, 5, <u>7</u>, 9, <u>11</u>, 13; SV: 3; Op: +2 **3.** 27, 22, 17, <u>12</u>, <u>7</u>, 2; SV: 27; Op: −5 **5.** 23, <u>34</u>, 45, 56, <u>67</u>, <u>78</u>; SV: 23; Op: +11
7. <u>2.3</u>, 2.7, 3.1, <u>3.5</u>, 3.9, <u>4.3</u>; SV: 2.3; Op: +0.4

9. a)

Minutes Spent Downhill Skiing	Total Daily Calories Burned
0	1,230
1	
2	
3	
4	

b) 6

c)

Minutes Spent Downhill Skiing	Total Daily Calories Burned
0	1,230
1	1,236
2	1,242
3	1,248
4	1,254

d) 1,470 calories

11. SV: 8; Op: +8; 40, 48, 56 **13.** SV: 9; Op: −4; −7, −11, −15

15. a) 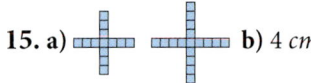 **b)** 4 cm
c) 12 cm; 20 cm **d)** SV: 4 Op: +8 **e)** 52 cm
17. Answers may vary. **19.**

x	$3(x-1)$	Output
−3	3(−3 − 1)	−12
0	3(0 − 1)	−3
2.2	3(2.2 − 1)	3.6
8	3(8 − 1)	21
21	3(21 − 1)	60

Lesson 2.2

1. SV: 10; Op: −2

x	y
0	10
1	8
2	6
3	4
4	2
5	0

3. SV: 2; Op: +0.5

x	y
0	2
1	2.5
2	3
3	3.5
4	4
5	4.5

5. SV: −1; Op: +2

x	y
0	−1
1	1
2	3
3	5
4	7
5	9

7. SV: 3 Op: none

x	y
0	3
1	3
2	3
3	3
4	3
5	3

9. a)

Time (seconds), x	Feet off the Ground, y
0	60
1	50
2	40
3	30
4	20
5	10
6	0

b) SV: 60; Op: −10
c) how tall the slide is
d) 6 seconds; See student work for explanation.

11. 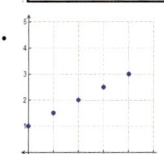 **13.** $12x - 18$ **15.** $-3x + 26$

Lesson 2.3

1. Answers may vary. **3.** Answers may vary.
5. a) SV: 799; Op: −70 **b)**

Years Passed	Value of Maggie's Laptop
0	$799
1	$729
2	$659
3	$589
4	$519
5	$449

c) 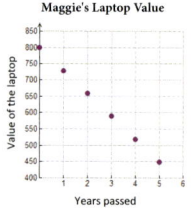 **d)** 12 years

7. a) SV: −61; Op: +7 **b)**

Minutes Hiking	Elevation
0	−61
1	−54
2	−47
3	−40
4	−33
5	−26
6	−19
7	−12
8	−5
9	2
10	9

c) **d)** approximately 9 minutes

9. a) SV: $1,000; Op: +$250 **b)** $5,500 **c)** $7,040
11. 60 minutes **13.** 5.8, 4.6, <u>3.4</u>, 2.2, <u>1</u>, <u>−0.2</u>; SV: 5.8; Op: −1.2
15. $\frac{1}{3}$, 1, $1\frac{2}{3}$, $2\frac{1}{3}$, 3, $3\frac{2}{3}$; SV: $\frac{1}{3}$ Op: $+\frac{2}{3}$ **17.** $x = -42$ **19.** $x = 8$
21. $x = -\frac{3}{2}$

Lesson 2.4

1. +2 bugs per day **3.** $50\frac{2}{3}$ steps per minute **5.** ROC: +4; SV: 5
7. ROC: +1; SV: −3 **9.** ROC: −2; SV: −1

11.

x	y
0	1
1	9
2	17
3	25
4	33
5	41

13.

x	y
−2	18
−1	13
0	8
1	3
3	−7
6	−22

15. a) 12 ants **b)** 30 ants **c)** 282 ants **17.** Answers may vary.
19. Answers may vary.

Lesson 2.5

1. $y = -6 + 8x$ **3.** $y = 7.1x$ **5.** $y = -10$ **7.** y-int = 4; ROC = +8; $y = 4 + 8x$ **9.** y-int = 6; ROC = +2; $y = 6 + 2x$ **11.** y-int = 7; ROC = +4; $y = 7 + 4x$ **13. a)** 2 problems per minute **b)** 12 **c)** $y = 12 + 2x$ **d)** minutes **e)** problems finished **15. a)** 65 **b)** 0 **c)** $y = 65$
17. y-int = −1; ROC = +3; $y = -1 + 3x$

x	y
3	8
4	11

19. y-int = 14; ROC = +11; $y = 14 + 11x$

x	y
5	69
7	91

21.

x	−2x + 7	Output
−2	−2(−2) +7	11
−1	−2(−1) +7	9
0	−2(0) +7	7
3	−2(3) +7	1
5	−2(5) +7	−3

23. 35 per hour **25.** 12.5 points per assignment

Lesson 2.6

1. y-int = 8; ROC = +2 **3.** y-int = −4; ROC = +1 **5.** y-int = 0; ROC = $-\frac{1}{4}$ **7.** y-int = −8; ROC = $\frac{2}{3}$ **9.** y-int = 2; ROC = $-\frac{4}{7}$

11.

x	x + 9	y
−4	−4 + 9	5
0	0 + 9	9
2	2 + 9	11
5	5 + 9	14
21	21 + 9	30

13.

x	y
−3	9
−1	3
6	−18
10	−30
20	−60

15.

x	y
−4	5
−3	5
0	5
1	5
5	5

17. She switched the order of the numbers in the ordered pairs. The correct ordered pairs are (2, 8), (3, 10) and (4, 12).

19. a)

x seconds	y meters run
10	68
25	170
40	272
60	408
100	680

b) ≈ 59 seconds; Answers may vary. **c)** 3,600 seconds **d)** 24,480 meters **e)** 15.3 miles **f)** not reasonable; one could not sprint that far **21.** $7\frac{1}{2}$ or 7.5 **23.** $x = -6$ **25.** $x = 3$ **27.** $x = -11$ **29.** $x = 3$

Lesson 2.7

1. $m = -3$ **3.** m = undefined **5.** $m = 0$ **7.** negative **9.** zero **11.** positive **13. a)**

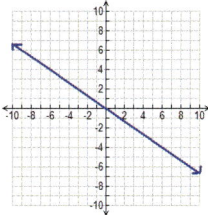

b) 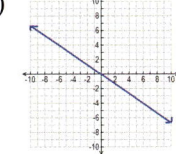 **c)** they are the same; yes; same as both

15. See student work.
17. Answers may vary.

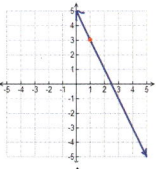

19. Answers may vary.
21. 21 feet
23. $\frac{2}{3}$
25. 9
27. undefined

Lesson 2.8

1. $\frac{3}{2}$ **3.** 0 **5.** undefined **7.** 0 **9.** −2 **11. a)** $\frac{4}{5}$ **b)** 0.8

c)

x	y
3	8
4	8.8
5	9.6
6	10.4
7	11.2
8	12

d) six; limitless

13. $-\frac{2}{5}$ or −0.4 **15. a)** $-\frac{1}{3}$ **b)** amount of pop draining out of the bottle per hour **c)** 6 hours; See student work for explanation. **17.** Answers may vary. **19.** F **21.** B **23.** G

Block 2 Review

1. 8, 1, −6, −13, −20, −27; SV: 8; Op: − 7
3. 7, 7.6, 8.2, 8.8, 9.4, 10; SV: 7; Op: +0.6
5. SV: 18; Op: −13; −34, −47, −60
7. a) 6   **b)** 10, 14 **c)** SV: 6; Op: +4 **d)** 30

9. SV: 2; Op: $+\frac{1}{2}$

x	y
0	2
1	$2\frac{1}{2}$
2	3
3	$3\frac{1}{2}$
4	4
5	$4\frac{1}{2}$
6	5

11. a) Answers may vary. **b)** Answers may vary.

13. a) SV: 3.35; Op: +0.40 **b)**

x	y
0	3.35
1	3.75
2	4.15
3	4.55
4	4.95
5	5.35

c) Predicted Gas Prices **d)** ≈ 17 years

15. 45 miles per hour **17.** SV: 2; ROC: +7
19. SV: 13; ROC: +4 **21.** $y = -2 + 6x$ **23.** $y = 1 + 3.8x$
25. ROC = +3.5; y-int = 9.9; $y = 9.9 + 3.5x$
27. a) $42 **b)** −$3.10 **c)** $y = 42 − 3.1x$ **d)** $26.50
29. ROC = −6; y-int = 7 **31.** ROC = 2; y-int = 7
33. ROC = 0; y-int = 8 **35.**
37. See student work; $m = \frac{1}{3}$

x	y
0	4
3	5
4	$5\frac{1}{3}$
6	6
11	$7\frac{2}{3}$

39. See student work; $m = -2$
41. Answers may vary.

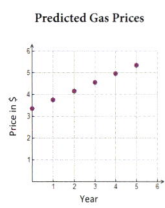

43. 9 feet **45.** $\frac{4}{3}$ **47.** 0 **49.** undefined **51. a)** rate of change; slope = −2 **b)** slope formula; slope = $-\frac{5}{4}$ **c)** slope triangle; slope = $\frac{3}{2}$ **53.** Answers may vary.

BLOCK 3

Lesson 3.1

1. **3.** **5.**

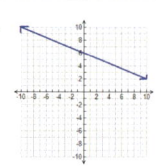 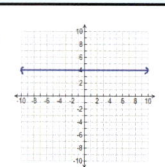

7. **9.** **11. a)**

x	y
−2	−1
0	2
2	5
4	8

b) **c)** Answers may vary.
13. Answers may vary.

15. a)

x	y
1	5.5
2	8
3	10.5
4	13
5	15.5

b) + 2.5 **c)** (0, 3) **d)** $y = 2.5x + 3$
e) 28 inches
17. C **19.** B **21.** undefined
23. 1 **25.** $-\frac{5}{3}$

Lesson 3.2

1. $m = \frac{3}{4}$; $b = 3$; $y = \frac{3}{4}x + 3$ **3.** $m = 1$; $b = -3$; $y = x - 3$
5. $m = -\frac{2}{3}$; $b = 0$; $y = -\frac{2}{3}x$ **7. a)** $y = 6x + 20$ **b)** $128
c) 32 months **9.** two points; you need to know the y-intercept and at least one other point **11. a)** $y = 3x + 5$
b) 29 inches **c)** 14 weeks **13. a)** $m \rightarrow y = \frac{1}{3}x + 2$; $n \rightarrow y = \frac{1}{3}x$; $p \rightarrow y = \frac{1}{3}x - 4$ **b)** same slope **c)** parallel **15.** false **17.** true
19. true **21.** about 27 minutes

Lesson 3.3

1. $y = \frac{6}{5}x + 8$ **3.** $y = x + 2$ **5. a)** $y = 45{,}000x + 4{,}600{,}000$
b) 5,050,000 people **7.** $y = -\frac{3}{4}x + 5$ **9.** $y = \frac{5}{2}x + 5$ **11.** $y = 8$
13. $y = 3x - 1$ **15.** $y = \frac{1}{3}x + 7$ **17.** $y = \frac{4}{5}x - 2$ **19.** disagree;
The equation of the line is not in slope-intercept form (for example; $x = 1$), but is still an equation for the line.
21. a) (2, 9) and (8, 12) **b)** $y = \frac{1}{2}x + 8$ **c)** 18 pounds; Answers may vary. **23. a)** $y = 3x + 14$ **b)** $14 **c)** cost per hour **d)** $26
25. $y = -\frac{1}{3}x + 4$ **27.** **29.**
31. $y = 0.3x + 3$

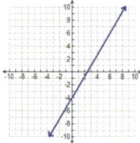 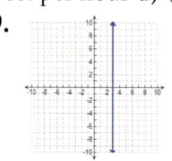

Lesson 3.4

1. C **3.** E **5.** F **7.** $y = 4x + 21$ **9.** $y = \frac{1}{3}x + 1$ **11.** $y = -\frac{1}{2}x - 1$
13. $y = \frac{3}{4}x + 7$ **15.** $y = 4x + 9$ **17.** $y = \frac{1}{3}(x - 9) + 1$;
See student work for explanation. **19.** $y = -3x + 10$
21. $y = -\frac{1}{2}x + 4$ **23.** $y = -5$ **25.** $y = -3x$

Lesson 3.5

1. $y = \frac{5}{2}x - 4$ 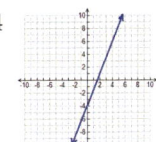 **3.** $y = \frac{1}{2}x + 3$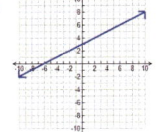

5. $y = \frac{4}{3}x - 2$ 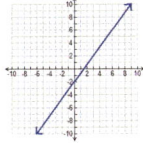 **7.** $x = 2$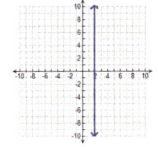

9. $y = 3x$ 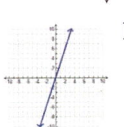 **11.** $y = x - 6$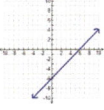

13. Answers may vary. See student work. **15.** yes **17.** yes
19. no **21.** No; she should have been charged $44.
23. $-3x + y = 4$ **25.** $y = -2x + 6$ **27.** $y = 2$ **29.** $y = 2x$
31. $y = \frac{5}{2}x - 4$ **33.** $y = \frac{1}{3}x - 1$

Lesson 3.6

1. Answers may vary. The inequality should have a < or > symbol. **3.** Clint is correct. A solid line shows that all points on the line will work in the equation or inequality.

5. 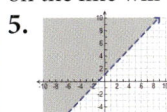 **7.** 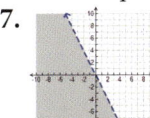 **9.**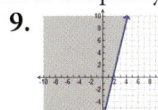

11. 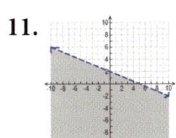 **13.** 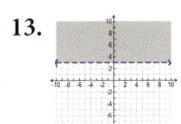 **15.** B **17.** D

19. 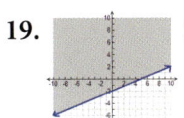 **21.** $y < -\frac{1}{4}x + 3$ **23.**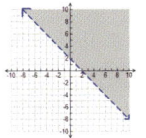

25. The point would not give information about which side of the line to shade. **27.** $y = \frac{2}{3}x + 1$ **29.** $y = 2x + 3$

Lesson 3.7

1.
x	y
-2	6
-1	3
0	2
1	3
2	6

3.
x	y
-2	-9
-1	-7
0	-5
1	-3
2	-1

5. exponential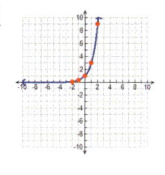

7. non-linear; exponential **9.** non-linear; inverse variation
11. Linear **13.** 640 bacteria **15.** A linear graph can have a constant and x- and y- variables that do not have exponents or are located in the denominator of a fraction.
17. 39 meters tall **19.** $x = 96$ **21.** $x = 6$ **23.** $x = 5$
25. 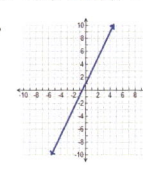 **27.** 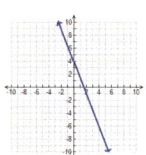 **29.**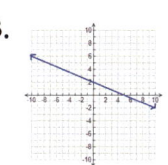

31. 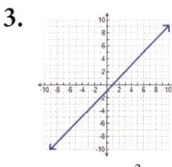 **33.**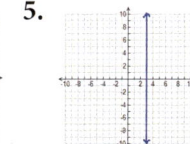

Block 3 Review

1. 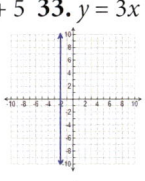 **3.** 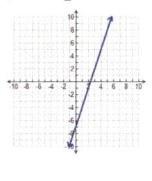 **5.**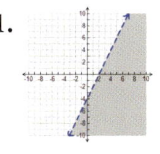

7. Answers may vary. **9.** $m = -\frac{3}{5}$; $b = -1$; $y = -\frac{3}{5}x - 1$
11. a) $y = 2x + 7$ **b)** 21 inches **c)** 11 weeks **13.** $y = -5x + 1$
15. $y = \frac{2}{5}x$ **17.** $y = \frac{1}{2}x - 2$ **19.** $y = \frac{5}{2}x + 5$ **21.** $y = \frac{1}{3}x + 2$
23. $y = -4$ **25. a)** (4, 32) and (10, 56) **b)** $y = 4x + 16$ **c)** 16
d) charge per hour **e)** $40 **27.** $y = -\frac{1}{2}x + 3$ **29.** $y = \frac{4}{5}x - 3$
31. $y = 3x + 5$ **33.** $y = 3x - 7$
35. $x = -2$

37. $y = 2x$ 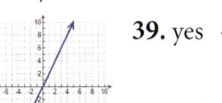 **39.** yes **41.**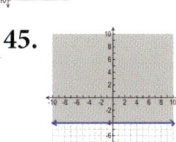

43. **45.** **47.** A **49.** D

51. non-linear; quadratic **53.** non-linear; exponential
55. linear **57. a)**
Year	$
1	4120
2	4243.60
3	4370.91

b) Non-linear because it does not increase by the same amount each time.

Selected Answers **265**

BLOCK 4

Lesson 4.1

1. same; infinite 3. parallel; none 5. same

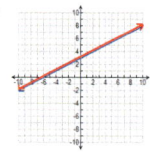

7. parallel; none 9. intersecting; one 11. same; infinite
13. yes; the equations have different slopes 15. their paths were the same 17. Answers may vary. 19. No; she gave different slopes and similar y-intercepts. She needed to give similar slopes and different y-intercepts. 21. false 23. true
25. $5x + 8$ 27. $4x - 8$ 29. $11x + 2y$

Lesson 4.2

1. a) put both in slope-intercept form b) graph c) find the point of intersection d) substitute back in and check
3. $(-1, 3)$ 5. no 7. yes 9. $(4, 5)$ 11. $(6, 4)$ 13. $(3, 3)$
15. $(6, 4)$ 17. no solution; parallel lines.
19. a) $y = 2x + 2y = 42$ b) $y = 2x + 3$ c)
d) length = 15 ft; width = 6 ft
21. Answers may vary.
23. intersecting; one 25. parallel; none
27. intersecting; one

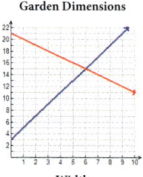

Lesson 4.3

1. $(1, 2)$

$y = 3x - 1$		$y = -2x + 4$	
x	y	x	y
0	−1	0	4
1	2	1	2
2	5	2	0
3	8	3	−2

3. they were getting further apart 5. $(1, 3)$ 7. $(-4, -2)$
9. $(0, 6)$ 11. a) $y = 15x + 100$
b) $y = -35x + 400$ c)

Carlos' Savings Account Balance

Months, x	Total Savings, y
0	$100
1	$115
2	$130
3	$145
4	$160
5	$175
6	$190
7	$205
8	$220

Ana's Savings Account Balance

Months, x	Total Savings, y
0	$400
1	$365
2	$330
3	$295
4	$260
5	$225
6	$190
7	$155
8	$120

d) after 6 months; They will both have $190. 13. 6 lawns; See student work for explanation. 15. $x = \frac{3}{2}$ 17. $x = -5$
19. $x = -2$

Lesson 4.4

1. x in #2 3. y in #2 5. sometimes the intersection point is not an integer or it can be hard to see on the graph
7. $y = 7 - 2x$ 9. $(-1, 2)$ 11. $(6, -2)$ 13. $(-7, 8)$ 15. $(-1, -6)$
17. $(-3, -1)$ 19. a) $x + y = 46$; $x - y = 12$ b) 29 & 17
21. $(4, 11)$; See student work for all three methods.
23. a) Tad: $6x + y = 67$; Timothy: $4x + 3y = 54$ b) $(10.5, 4)$
c) $10.50 for a can of paint; $4.00 for a brush
25. intersecting; one 27. intersecting; one 29. parallel; none

Lesson 4.5

1. See student work. (e.g., $x + 4y = 23$
$-x + y = 2$)
3. See student work. (e.g., $5x + 2y = 6$
$-9x - 2y = -22$)
5. See student work. (e.g., $9x + 6y = 24$
$4x - 6y = -24$)
7. She did not make it so the x- or y-coefficients were opposites. 9. $(-2, 3)$ 11. $(5, -3)$ 13. $\left(0, -\frac{1}{2}\right)$ 15. $\left(\frac{3}{4}, 1\right)$
17. a) $3x + y = 29$; $x + 2y = 18$ b) one block is 8 and the other is 5 19. $(5, 2)$ 21. $(4, 0)$ 23. $(3, -1)$

Lesson 4.6

1. elimination. The x's are already opposites and in standard form. 3. graphing. Both equations are in slope-intercept form. 5. elimination; $(4, -2)$ 7. elimination; $(3, 3)$
9. substitution; $(2, 4)$ 11. elimination; $(-2, -6)$
13. substitution; $(1, 2.5)$

Lesson 4.7

1. x = weeks; y = total savings; Manny: $y = 85x + 200$; Susan: $y = 100x + 95$ 3. w = width of the garden; l = length of the garden; $2w + 2l = 184$; $w = 2l$ 5. x = number of snaps; y = total cost; old machine: $y = 0.12x + 1{,}100$ new machine: $y = 0.09x + 1520$; $(14000, 2780)$; The company would need to attach at least 14,000 snaps to make the new machine worthwhile. 7. x = family A's total; y = family B's total; $x + y = 1{,}640$; $x = 2y + 182$; $(1154, 486)$; One family made $1,154 and one family made $486.
9. x = price of a gallon of ice cream; y = price of a container of strawberries; Jeremiah: $3x + 4y = 19.50$; Gary: $5x + 2y = 22$; $(3.5, 2.25)$; A gallon of ice cream costs $3.50; a container of strawberries costs $2.25.
11. x = number of hours; y = total miles traveled; Nancy: $y = 50x + 72$; Pedro: $y = 62x$; $(6, 372)$; He will catch up with her after 6 hours. 13.   15.
17.   19. $x \leq 2$

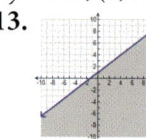 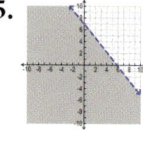

Lesson 4.8

1. Solutions are represented by the double-shaded region.

3. 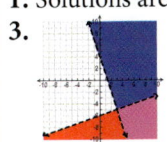 **5.** 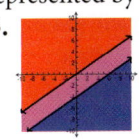 **7.**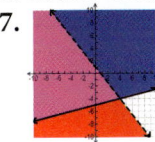

9. 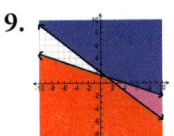 **11.** A **13.** B

15. 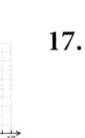 **17.**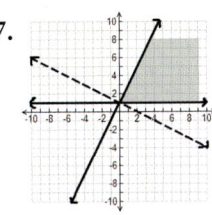

19. $y > 0$ and $x > 0$; See student work for explanation.
21. Answers may vary. **23.** 21 square units; See student work. **25.** yes **27.** $(-6, -9)$ **29.** $(6, 2)$ **31.** $(6, -3)$

Lesson 4.9

1. terminating decimal **3.** repeating decimal **5.** repeating decimal **7.** $\frac{2}{11}$ **9.** $\frac{1}{6}$ **11.** $\frac{1}{9}$ **13.** $\frac{2}{9}$ **15.** $\frac{5}{33}$ **17.** $\frac{11}{90}$ **19.** $\frac{46}{111}$ **21.** $\frac{6}{37}$ **23.** Decimals are all based on multiples of ten.
25. a) $9\frac{13}{30}, 9\frac{4}{9}, 9.\overline{45}$ **b)** Anna **27.** Answers may vary.
29. a) 4 nights **b)** $570 **31. a)** $9 **b)** $6 **c)** $54

Block 4 Review

1. intersecting; one solution **3.** parallel; no solutions
5. parallel; no solutions **7.** yes **9.** yes **11.** $(-2, 0)$ **13.** $(6, -2)$
15. $(-4, -9)$ **17.** $(-2, 9)$

$y = -5 + x$		$y = 3x + 3$	
x	y	x	y
−5	−10	−5	−12
−4	−9	−4	−9
−3	−8	−3	−6
−2	−7	−2	−3

19. a) $y = 400 - 20x$ **b)** $y = 100 + 40x$
c)

Evan's Balance		Lisa's Balance	
Months, x	Total Savings, y	Months, x	Total Savings, y
0	$400	0	$100
1	$380	1	$140
2	$360	2	$180
3	$340	3	$220
4	$320	4	$260
5	$300	5	$300
6	$280	6	$340

d) after 5 months they will both have $300 **21.** $(-3, 0)$
23. a) $x + y = 79$ represents the total number of Mr. James and Mr. Peters' cows; $x = 3y - 5$ represents 5 less than three times Mr. Peters' amount **b)** $(58, 21)$; Mr. James = 58 cows; Mr. Peters = 21 cows **25.** See student work. (e.g., $6x - 2y = 6$
$6x + 2y = 9$)
27. $(-1, 6)$ **29.** $(1.5, 4)$ **31.** elimination; $(4, -2)$
33. graphing; $(3, -4)$ **35.** substitution; $(-6, 0)$
37. x = number of hours; y = total cost of rental; A: $y = 2.5x + 8$; B: $y = x + 14$; $(4, 18)$; after 4 hours they are the same amount; rentals are $18 for both companies after 4 hours **39.** x = months passed; y = total amount in savings account; Jamal: $y = 24x + 46$; Emily: $y = -15x + 319$; $(7, 214)$; after 7 months they will both have $214 in their accounts
41. **43.** 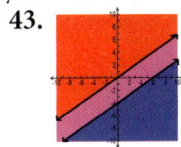 **45.**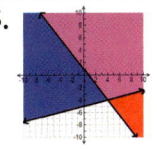

47. Agree; see student work. **49.** repeating **51.** $\frac{5}{9}$ **53.** $\frac{23}{99}$
55. $1\frac{8}{15}$ **57.** Answers may vary.

BLOCK 5

Lesson 5.1

1. Answers may vary. Ordered pairs make up the points in a scatter plot. **3.**

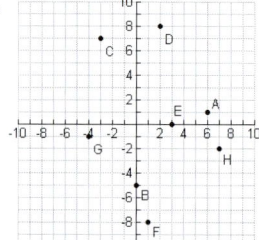

5. a) horse **b)** giant tortoise **c)** mouse **d)** Answers may vary. **e)** Answers may vary. **7.** negative; As the price of gas goes up fewer people are likely to buy fuel there. **9.** negative; If you brush your teeth you are less likely to have cavities.
11. a) 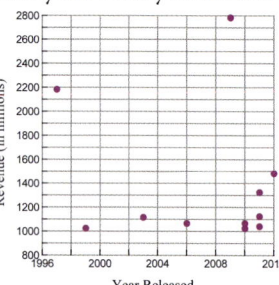 **b)** There is no correlation.

13. (1, 4) **15. a)** $l = 2w + 4$, $2l + 2w = 104$; l = maximum length of the pen, w = maximum width of the pen; The maximum width of the pen is 16 feet and the maximum length is 36 feet.

Lesson 5.2

1. follows the pattern of the data and has about half the points below the line and half the points above the line **3.** yes **5. a)** 7 **b)** 7 **c)** Answers may vary. **d)** no; not a linear pattern **7.** 10 **9.** 47.5 **11. a)** negative correlation **b)** line A **c)** about 2,500 square feet **d)** about 10 students **13.** Answers may vary. **15.** no correlation **17.** negative correlation

Lesson 5.3

1. a) minimum, median, maximum **b)** Q1 and Q3 are wrong; 19 ~ 23 ~ 26 ~ 28 ~ 30 **c)** Range = 11, IQR = 5 **d)** 25% **e)** 26 **3.** 40 ~ 43 ~ 46 ~ 54 ~ 60 **5.** 52 ~ 59.5 ~ 74 ~ 77 ~ 79 **7.** −9 ~ −5 ~ 0 ~ 7 ~ 10 **9.** 9 **11. a)** 2.6 pounds **b)** 75% **c)** 2.4 and 2.9 **13.** 2, 11, 15 **15.** 14, 29 **17.** no; The line does not follow the direction of the data well. **19.** 7.5 **21.** 9 **23.** $3.50 for a single bag of dried fruit

Lesson 5.4

1. Answers may vary. Q-points gives a process for finding a line of best fit so that everyone will get the same answer. In Lesson 5.2, two students may disagree on how to best draw their line of best fit. **3.** She should choose the points that follow the pattern of the data. **5.** (3, 6), (10, 6), (10, 17), (3, 17) **7. a)** (1, 9), (2, 8), (4, 18), (6, 20), (6, 23), (7, 30), (8, 29), (10, 37) **b)** 1 ~ 3 ~ 6 ~ 7.5 ~ 10
c) 8 ~ 13.5 ~ 21.5 ~ 29.5 ~ 37 **d)**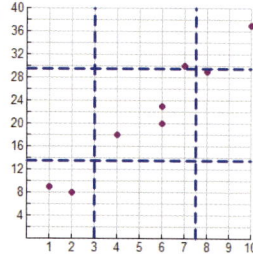
e) (3, 13.5) and (7.5, 29.5)
f) no; They are quartile points and do not necessarily have to be points in the data set.

9. $y = 1.5x + 2$ **11.** $y = -x + 13$ **13. a)** yes; negative correlation **b)** Answers may vary. No; if the graph continued in the same way, a 50-year-old would have 0 or negative meals per day which is not reasonable. **15.** $x = 17$
17. $x = 1.4$

Lesson 5.5

1. 46 **3.** 17 **5. a)** 40 **b)** 21 students **c)** the year 2035 **d)** Answers may vary. Classes with an average of 15 students may or may not be reasonable. **7. a)** negative correlation **b)** (1936, 10.4) and (1996, 9.9) **c)** 9.5 seconds **d)** the year 2233; Answers may vary. It may not be possible for a person to ever run 100 meters in 8 seconds. **9. a)**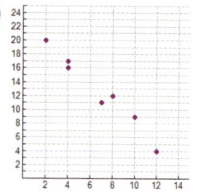
b) (4, 9), (10, 9), (4, 17), (10, 17)
11. $y = -3x + 5$

Lesson 5.6

1. A sample is a small portion of an entire population.
3. a) Brad surveyed his friends; they are not likely to be a random sample. **b)** Math was one of only three options. **c)** Data could be skewed by the fact that some students had already gone to bed. Also, Brad's friends do not give a random sample of the number of A's that students have. **d)** Only students who tend to stay up late at night would be responding. **e)** Answers may vary. Brad could attempt to get a more random sample of students to participate. **5.** Bias is in the "who"; being the boss may affect how people respond to the survey. **7.** Bias is in the "who"; friends do not constitute a random sampling. **9.** Bias is in the "when"; early arriving students may lead to misleading data. **11.** Answers may vary. **13.** 23 **15.** −8 **17. a)** positive correlation **b)** (7, 36), (17, 48) **c)** $y = 1.2x + 27.6$

Lesson 5.7

1. The last statement should be: There were 120 people surveyed. **3. a)** 120 **b)** 70 people were taller than 6 feet and played basketball; 15 people were taller than 6 feet and did not play basketball; 5 people were not taller than 6 feet and played basketball; 30 people were not taller than 6 feet and did not play basketball

5. Likes to Fly

		Y	N
Owns a Passport	Y	60	15
	N	10	5

7. Runs ≥ 4

		Y	N
Win	Y	45	15
	N	8	12

9. A relative frequency is the ratio of a number to the entire sample surveyed. A conditional frequency is the ratio of a number to the total surveyed in one category.

11. a) Play Tennis

		Y	N
Play Golf	Y	0.2	0.3
	N	0.3	0.2

b) Answers may vary.

13. play golf and play tennis = 0.4; play golf and don't play tennis = 0.6; No, it is more likely if a person plays tennis she will play golf. **15.** Answers may vary (e.g., using relative frequencies, most people play neither (0.4)). **17.** Measuring only the heights of the boys does not give a random sample. Answers may vary. Instead, pick a random 8th grade class and measure all students (boys and girls).

Block 5 Review

1. $A(0, -5)$; $C(-7, -4)$; $E(3, -6)$; $R(6, 0)$; $S(7, 2)$; $T(-4, 8)$
3. negative **5.** no correlation **7.** follows the pattern of the data and has about half the points above the line and half below **9.** yes **11.** 82.5 **13.** 22 ~ 29 ~ 32 ~ 36 ~ 40
15. 4 ~ 7.5 ~ 15 ~ 19.5 ~ 25 **17.** 14 **19. a)** 25% **b)** 50%
c) 50% **21.** (3, 27) and (8, 12); $y = -3x + 36$
23. a) (2, 8), (4, 12), (5, 16), (5, 18), (6, 22), (7, 22), (8, 26), (10, 30) **b)** 2 ~ 4.5 ~ 5.5 ~ 7.5 ~ 10
c) 8 ~ 14 ~ 20 ~ 24 ~ 30 **d)**
e) (4.5, 14) and (7.5, 24)

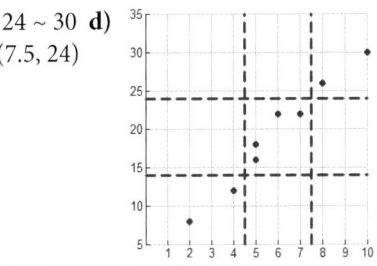

25. 21 **27. a)** $98 **b)** approximately 75 pieces
29. asking only your friends is not a random sample; Answers may vary. Instead, ask random students in the cafeteria at lunchtime. **31.** Mail-in questionnaires makes responding optional. Answers may vary. Instead, ask a random sampling of people leaving the theater to rate the movie. **33.** Employees may not be honest with the boss who just gave them a raise. Answers may vary. Instead, have random employees take an anonymous survey. **35. a)** 80
b) There are forty students who ride a bike and a skateboard. There are thirty students who ride a bike, but do not ride a skateboard. There are five students who did not ride a bike, but did ride a skateboard. There are five students who did not ride a bike or a skateboard.

37. Ride the Bus

		Y	N
Late to 1st Period	Y	20	15
	N	10	15

39. a) Brought Lunch

		Y	N
Purchased Lunch	Y	0.15	0.4
	N	0.4	0.05

Answers may vary.

b) The conditional frequency for a person who brought lunch to school and purchased their lunch is $0.\overline{27}$. The conditional frequency for a person who brought lunch to school and did not purchase their lunch is $0.\overline{72}$.

Selected Answers **269**

INDEX

A
Absolute value, 3

Algebraic expression, 6
   coefficient, 10
   constant, 11
   evaluating, 6
   simplifying, 11
   term, 10

Axes, 32

B

C
Career Focus
   Actuary, 224
   Accountant, 180
   Insurance Agency Owner, 43
   Newspaper Editor, 92
   Retail Manager, 134

Coefficient, 10

Constant, 10

Continuous graph, 97

Converting linear equations, 111
   Explore! One of These Things, 112
   and graphing, 115
      Explore! Match Me, 115

Coordinate plane, 51
   axes, 51
   ordered pair, 51
   origin, 51
   quadrants, 51
   scatter plot, 183
   x-axis, 51
   y-axis, 51

D
Dependent variable, 58

Discrete graph, 97

Distributive Property, 10

E
Elimination method, 154

Equations, 7
   multi-step, 25
   one-step, 15
   two-step, 20

Equations, 7
   writing from statements, 17

Equation mats, 15

Equivalent, 112

Equivalent expressions, 11
   Explore! Match Them Up, 12

Evaluating expressions, 6

Explore!
   A Trip on I-70, 151
   Addition and Subtraction
      Equations, 15
   At the Movies, 161
   Caloric Recursive Routines, 46
   Find that Formula, 82
   Find the Equation, 100
   Larry's Landscaping, 145
   Linear Qualities, 73
   Match Me, 115
   Match Them Up, 12
   Modeling with Equations, 67
   Multi-Step Equations, 26
   Non-Linear Curves, 125
   One of These Things, 112
   Saving and Spending, 56
   Triangle lines, 107
   Types of Systems, 137
   What's Easiest?, 158
   What Works?, 30

Exponential functions, 125

F

G
Graphing
   continuous lines, 97
   discrete, 97
   from slope-intercept form, 95

H

I
Independent variable, 58

Input-output tables, 73

Explore! Linear Qualities, 73

Inverse operations, 16

Inverse variation functions, 125

J

K

L
Like terms, 11

Linear plots, 51

Linear equations, 67
   converting to, 111
   Explore! Modeling with Equations, 67
   from graphs, 100
      Explore! Find the Equation, 100
   from key information, 106
      Explore! Triangle Lines, 107
   from recursive routines, 67
   point-slope form, 111
   solutions to, 30
   standard form, 111

M
Multi-step equations, 25
   Explore! Multi-Step Equations, 26

N
Non-linear functions, 125
   Explore! Non-Linear Curves, 125
   exponential functions, 125
   inverse variation functions, 126
   parent graphs, 125
   quadratic functions, 125

O
One-step equations, 15
   Explore! Addition and Subtraction
      Equations, 15

Order of operations
   with four operations, 3

Ordered pair, 51

Origin, 51

P
Parallel, 138

Parent graphs, 125

Point-slope form, 111

Properties of equality, 15

Q
Quadrants, 32

Quadratic functions, 125

R
Rate of change, 61

Recursive routine, 46
   applications, 56
      Explore! Saving and Spending, 56
      Explore! Caloric Recursive Routines, 46
   start value, 46
   using a calculator, 48
   writing linear equations, 67

Recursive sequences, 46
   start value, 46

S
Scatter plot, 183

Simplify
   algebraic expressions, 11

Slope, 77
   negative, 77
   positive, 77
   undefined, 77
   zero, 77

Slope formula, 82
   Explore! Find that Formula, 82

Slope-intercept form, 95
   graphing from, 95

Slope triangles, 77

Solution to a system of equations, 137
   by graphing, 141

Standard form, 111

Start value, 46

Substitution method, 150
   Explore! A Trip on I-70, 151

System of linear equations, 137
   applications, 161
      Explore! At the Movies, 161
   choosing a method to solve, 159
      Explore! What's Easiest?, 158
   elimination method, 154
      Explore! Types of Systems, 137

System of linear equations, 137
   graphing, 141
   solution to, 137
   substitution method, 150
      Explore! A Trip on I-70, 151
   types of, 137
   using tables to solve, 146
      Explore! Larry's Landscaping, 145

T
Term, 10
   like terms, 11

Two-step equations, 20

Types of systems of equations, 137

U
Undefined slope, 77

V

W

X
x-Axis, 51

Y
y-Axis, 51

y-Intercept, 68

Z
Zero pair, 15

Zero slope, 77

PROBLEM-SOLVING

UNDERSTAND THE SITUATION

- Read then re-read the problem.
- Identify what the problem is asking you to find.
- Locate the key information.

PLAN YOUR APPROACH

Choose a strategy to solve the problem:

- Guess, check and revise
- Use an equation
- Use a formula
- Draw a picture
- Draw a graph
- Make a table
- Make a chart
- Make a list
- Look for patterns
- Compute or simplify

STOP AND THINK

- Did you answer the question that was asked?
- Does your answer make sense?
- Does your answer have the correct units?
- Look back over your work and correct any mistakes.

SOLVE THE PROBLEM

- Use your strategy to solve the problem.
- Show all work.

DEFEND YOUR ANSWER

Show that your answer is correct by doing one of the following:

- Use a second strategy to get the same answer.
- Verify that your first calculations are accurate by repeating your process.

ANSWER THE QUESTION

- State your answer in a complete sentence.
- Include the appropriate units.

SYMBOLS

Algebra and Number Operations

SYMBOL	MEANING
+	Plus or positive
−	Minus or negative
$5 \times n,\ 5 \cdot n,\ 5n,\ 5(n)$	Times (multiplication)
$3 \div 4,\ 4\overline{)3},\ \frac{3}{4}$	Divided by (division)
=	Is equal to
≈	Is approximately
<	Is less than
>	Is greater than
%	Percent
$a : b$ or $\frac{a}{b}$	Ratio of a to b
$5.\overline{2}$	Repeating decimal (5.222…)
≥	Is greater than or equal to
≤	Is less than or equal to
x^n	The nth power of x
(a, b)	Ordered pair where a is the x-coordinate and b is the y-coordinate
±	Plus or minus
$\sqrt{x}$	Square root of x
≠	Not equal to
$x \stackrel{?}{=} y$	Is x equal to y?
$|x|$	Absolute value of x
$P(A)$	Probability of event A

Geometry and Measurement

SYMBOL	MEANING
≅	Is congruent to
~	Is similar to
∠	Angle
$m\angle$	Measure of angle
$\triangle ABC$	Triangle ABC
$\overline{AB}$	Line segment AB
$\overrightarrow{AB}$	Ray AB
AB	Length of AB
π	Pi (approximately $\frac{22}{7}$ or 3.14)
°	Degree